An Attempt to Restore Classical Physics

Dr. Paul E. Rowe

iUniverse, Inc.
New York   Bloomington

An Attempt to Restore Classical Physics

iUniverse books may be ordered through booksellers or by contacting:

iUniverse
1663 Liberty Drive
Bloomington, IN 47403
www.iuniverse.com
1-800-Authors (1-800-288-4677)

Because of the dynamic nature of the Internet, any Web addresses or links contained in this book may have changed since publication and may no longer be valid. The views expressed in this work are solely those of the author and do not necessarily reflect the views of the publisher, and the publisher hereby disclaims any responsibility for them.

ISBN: 978-1-4502-2233-4 (sc)
ISBN: 978-1-4502-2235-8 (ebook)
ISBN: 978-1-4502-2234-1 (dj)

Printed in the United States of America

iUniverse rev. date: 4/16/2010

ACKNOWEDGMENTS

The Author wishes to thank the following:

H. W. Milnes, editor of "The Toth- Maatian Review" for publishing my early articles.

Eugene Mallove, editor of "Infinite Energy", for publishing many of my articles prior to his murder in 2004.

Sepp Hasslberger for including many of my article in his blog.

Professor Harold Edgerton for commenting on my first article and encouraging me to continue this effort.

Professor Philip Morrison for discussing my ideas, even though he didn't agree with them.

Professor Linus Pauling for reading and commenting on two of my letters.

Mort Spears and Jay Schleiker for comparing their ideas and mine in many stimulating conversations.

Emmet Harrison, Lou Watkins, Bob and Georgia Kelley, Paul and Marcia Gaudette, Bob Wittenaur and William Rowe for reading and commenting on my articles.

My wife Kate for everything and especially for putting up with me all these years.

CONTENTS

INTRODUCTION

Based on the results of my experiments and extensive searches in the stacks of the MIT Science Library, I believe that the aether of classical physics is a concentrated matrix of protons and unpaired electrons. It may also be the "Dark Matter" for which scientist are searching. It may be Bose-Einstein condensed hydrogen. This book attempts to explain many scientific observations, on this basis.

BACKGROUND

During World War II, German shells were found to be much more powerful than similar American shells. Analysis of unexploded German shells revealed they contained mixtures of aluminum flake and high explosive. Experiments revealed that the presence of aluminum increased the time that pressure from the explosion persisted, when the explosion took place in air. The effect was called "after burning". When exploded in vacuum, aluminum reduced the power of the shells. Some experimenters thought that molten aluminum, dispersed by the explosion, combined with the air to form aluminum oxide and that this reaction produced energy that heated air in the target.

When I detonated shells containing only high explosive, in vacuum, I obtained the amount of gas expected based on the chemical composition of the explosive. On, similarly, detonating shells containing explosive- aluminum mixtures, I obtained much more gas than was theoretically possible, based on the known ingredients in the shell. Increasing the aluminum concentration above the concentration required to completely react with the explosive increased the amount of gas produced.

I tested simpler mixtures. Combustion, in vacuum, of highly purified aluminum powder and CP (Chemically Pure) cupric oxide produced the gas most readily. Again, much more than the calculated amount of gas was produced. Increasing the percentage of aluminum produced more gas. The quantity of gas produced from 10 grams of mixture produced volumes of gas easily measured by a mercury

3

manometer. The pressure obtained after sparking mixtures of this gas and air convinced me that the original gas was hydrogen. Could the hydrogen have been produced from vacuum, when very hot molten aluminum droplets were dispersed in vacuum? If so, the hydrogen produced during explosions in air would be expected to react with oxygen of the air to produce heat and increase the power of the explosion. Could the sparks observed while grinding some metals, in air, be caused by reaction of similarly produced hydrogen with the oxygen in the air?

Is there something in vacuum which can be converted into hydrogen, under the proper conditions?

This book is my effort to explain the results of these and other experiments.

DISCUSSION

James Clerk Maxwell developed equations that predict various properties of electromagnetic radiation based on the following assumptions:

- A medium for light transfer is present in vacuum.
- The medium is made up of touching material particles
- The medium has specific magnetic and dielectric properties.

Much of Albert Einstein's work is based on Maxwell's equations and, therefore, on Maxwell's assumptions.

The following are attempts to explain observed phenomena based on the existence of an aether of protons and unpaired electrons.

Magnetic Properties of Vacuum

The forces encountered when manipulating separated bar magnets, are strong evidence that the space between the magnets is paramagnetic. This suggests that space contains unpaired electrons. I have produced surprisingly large quantities of hydrogen gas by combusting aluminum powder- cupric oxide mixtures in vacuum. Many highly respected experimenters have reported the surprising appearance of hydrogen gas in low pressure electrical discharge tubes . Clarence Skinner produced hydrogen gas in electric discharges in low pressure helium. The initial rate of production followed Faraday's laws of electrolysis. The medium may be a matrix of protons and unpaired electrons, much as molten common salt, where no sodium ion touches another sodium ion and no chloride ion touches another chloride ion.

DIELECTRIC CONSTANT OF VACUUM

Vacuum has a definite dielectric constant. Aether between separated, oppositely charged metal plates would be distorted because aether electrons would be attracted toward the positive plate and aether protons toward the negative plate. Void cannot have a dielectric constant.

ELECTROMAGNETIC RADIATION

When a DC current passes through a straight wire, a compass needle near the wire points perpendicular to the wire. It remains so oriented until the current is stopped. This suggests that moving electrons on the wire cause aether electrons in the vicinity of the wire to orient. When the current is reversed, the needle points in the opposite direction. At extremely low frequencies, the orientation of the needle changes with frequency. At higher frequencies, the needle of the compass cannot reorient fast enough because of its inertia. Aether electrons have very little inertia and continue to orient with changing current direction, even at extremely high frequencies. (Could Planck's constant be related to the inertia of the electron to the type of rotation required?). Let us consider a half-wave dipole antenna operating at a FM frequency. I understand that the current moves along the surface of such an antenna at the speed of light. According to Huygens' principle, each orienting aether electron passes all of its energy to a neighboring aether electron, etc. This effect moves away from the antenna at the speed of light. As the direction of the current changes, a line of energy starts forming at one end of the antenna. Just as the direction of the current changes again, this line of energy leaves the other end of the antenna. The result is a line of energy moving through the ether at the speed of light (a photon?). This line is in a plane of the antenna and at an angle of 45 degrees to the antenna. When the direction of the current changes again, a similar line of energy (photon?) is produced. This photon is also in a plane of the antenna but at an angle of 135 degrees to the antenna. Its aether electrons are orienting oppositely. What we consider one wave in the antenna may produce a pair of photons. Each photon might be considered a mirror image of the previous photon. At any instant, an integer number of active electrons are involved in a photon. There are no fractional electrons. For this reason, only specific amounts of energy can be carried as photons. This may be the basis for quantum mechanics. Vibrating chemical bonds produce a similar effect . A photon is produced as the separation increases and then, a mirror image photon is produced as the separation decreases.

GRAVITY

Addition of the masses of any material's protons and neutrons yields values close of the mass of that material. If one assumes that a neutron is a combination of a proton and an electron, the mass of any material is very close to the masses of the protons it contains. This led me to suspect there is an attraction between protons that becomes dominant when the electrical forces of the protons are cancelled by electrons. Could this force be gravity?

Lord Rutherford's group passed alpha particles (helium nuclei) through gold leaf and, based on scattering of alpha particles, concluded that that the nucleus of gold was extremely tiny compared to the spaces between the gold nuclei in the gold leaf. More recent experiments indicate that the diameter of any nucleus is about 1/10000 times the diameter of its atom. Since a nucleus contains protons and electrons, their diameters are, likely, smaller than that of all nuclei, with the possible exception of hydrogen.

Bose-Einstein condensed atoms are predicted to have zero viscosity. Condensates of elements heavier than hydrogen (lithium, sodium and rubidium) transfer light at slower velocities than space. Bose-Einstein condensed hydrogen might be expected to transfer light much faster than condensates of heavier atoms. The distance between nuclei in solids and liquids are vast gaps to Bose-Einstein condensed hydrogen. Any material would be expected to move through the proposed matrix with no resistance. A sieve moves through water similarly, but with some resistance.

Picture a nucleus as a positively charged sphere. Such a sphere in the proposed matrix would attract aether electrons of the matrix and repel protons of the matrix. Since the atom is electrically neutral the distortion of the matrix must balance the charge of the nucleus. If so, the exterior of the atom has a positive charge. Could the distorted matrix, in the vicinity of the nucleus, be required to give the atom a neutral charge and help stabilize the nucleus?

Rutherford's calculations are based on the size of the helium nucleus. A helium atom is vastly larger. Perhaps, as nuclei move through the aether, the aether adapts as the nucleus moves through it. The aether may have a problem reacting fast enough, when the speed of a nucleus approaches the speed of light. Perhaps only the nuclei move and the aether adapts to that movement.

THE NATURE OF THE CHEMICAL BOND?

The spaces between nuclei in molecules are vast gaps compared to the diameter of each nucleus. I suspect the spaces are filled with the proposed aether as are all spaces in the knowable universe. Picture two widely separated atoms. As the atoms approach each other, the gravitational attraction increases. At some distance the electrical repulsion of the positive outer surfaces of the atoms balances the gravitational attraction. Could this be the chemical bond length?

STELLAR ABERRATION

When astronomers on earth try to fix the position of stars in space, they correct their measurement for known effects like bending of light as it enters the earth atmosphere. In spite of these corrections, the position of any star seems to change slightly as the earth rotates around the sun. In the northern hemisphere the North Star appears to move in a small circle. The apparent position depends on the date of the year on earth. No one believes the North Star's actual position is affected by the date on the earth. This effect is called stellar aberration.

When a light beam passes close to a massive object, the beam is bent towards that object. According to Maxwell's equations, this indicates that the aether in the vicinity of the object has a higher dielectric constant and/or magnetic permeability than the aether of free space. It also, suggests the aether in the vicinity of the object is denser than the aether of free space. This is expected since the density of the massive object is greater than the density of free space. Since air tends to move along with the surface of the earth, we expect the aether near the surface of the earth to tend to move with the surface of the earth, as the earth spins. If so, the results of the Michelson-Morley interferometer experiments are just as should be expected. Should we consider the Coriolis effect?

If the portion of the aether in the vicinity of the earth moves with the earth as it orbits around the sun, at about 26 miles per second, light from a star bends as it approaches the earth. A similar effect occurs when a sound wave encounters wind. In the case of starlight, the direction of bending, relative to the star, reverses every six months. In the case of the North Star, one would expect the deviation to be nearly circular. The observed position should be the same on the same date each year.

EINSTEIN'S TWIN PARADOX

As an airplane approaches the sound barrier, more and more energy is required to impart a given acceleration. The mass of the airplane is believed to remain constant. A greater force is required because the air through which the airplane is traveling increases its resistance to increased velocity. The medium for light may produce similar effects, as a rocket ship approaches the "light barrier". This is possible if the energy increase is in the material and not transferred to the aether. Huygens believed that matter is an open mesh through which the aether easily passes. However, as a nuceus approaches the speed of light, aether particles may have increasing difficulty adapting to the nucleus. If so, the energy required for imparting a given acceleration may increases with increasing velocity.

If one defines a unit of mass based on a standard that is at rest, relative to the aether, in its vicinity, one expects greater energy to be required to accelerate an object as the object approaches the speed of light. By this definition of mass, the force required to accelerate an object increases as the object approaches the speed of light, its mass does not change as Einstein has proposed.

If materials that make up a rocket ship are porous to the ether, a pendulum clock in a rocket ship moves through the ether at a similar rate as the rocket ship, itself. As the clock approached the speed of light, the period of the pendulum would decrease and time, as measured by the clock, would decrease. The same effect would occur with modern time devices that are based on the periods of certain bond vibrations.

In the twin paradox, one twin remains on earth while the other twin rockets away at close to the speed of light, turns around and then returns at a similar speed. The speeding twin's clock would run slower in both directions of flight, since in each case, it is moving rapidly through the aether. The earthbound twin's clock is comparatively still, relatively to the aether in its vicinity, and would indicate the passage of more time. At the speed of light, the pendulum and the vibration might stop.

I prefer a definition of time based on a clock that is stationary, relative to the aether, in its vicinity. In order to determine the correct time, a factor, based on the speed at which a clock is moving through the aether (or, if you prefer, the aether is moving through the clock) would be applied to the time it registers. Time would, then, be more nearly universal and independent of velocity.

If I am correct, the vibrations of the chemical bonds in the speeding twin's body would be slowed dramatically. This would slow all the chemical reactions in his body. If he survives, he may well appear younger than the stationary twin.

STAR FORMATION

There is evidence that hydrogen gas is being formed in various parts of the universe. The hydrogen gas may be pulled together by gravity until the agglomeration attains sufficient size, density and temperature to initiate a fusion reaction and turn the agglomeration into a star. Where does all this hydrogen come from? If experimenters can convert the aether into hydrogen in their laboratories, should one be surprised that nature performs the same conversion in deep space?

ASTRONOMY

I have no clue as to whether the quantity of ether is finite. If it is finite, the knowable universe might be expected to be a sphere surrounded by void. The ether- void interface would be a nearly perfect mirror and there would, likely, be no way of knowing whether observed radiation had been reflected from that interface. Perhaps, some of the radiation believed to be from great distances originated in our own galaxy and has been reflected back to us."

THE RED SHIFT

Spectral lines in light from distant stars are shifted toward lower frequencies, that is, to lower energy. The greater the distance light has travelled, the greater the shift in frequency. It is assumed that the stars are moving away from us and the farther they are from us, the faster they are moving relative to the earth. Most astronomers conclude that the universe is expanding and that there was a "Big Bang" that that initiated the expansion. If the volume of the aether is the volume of the universe, expansion requires an increase in the separation between aether particles or a great increase in the number of aether particles. Either prospect seems unlikely. An alternate conclusion is that objects in space are, on the average, moving away from each other and filling more and more of the aether. The shift in light to lower frequencies can be explained, just as readily, if one assumes that light slowly loses energy as it travels great distances through the aether, or if some energy is lost when light is reflected from the aether –void barrier, of a less than infinite universe. If, as I suspect, hydrogen is produced in space, the energy required may be removed from photons traveling through space. The hydrogen and other particles may remove energy from photons resulting in the observed red shift. If this is the case, the universe may not be expanding.

PLANCK'S CONSTANT

Planck's constant is the ratio of the energy of the radiation to the frequency of the radiation. It may actually be the ratio of the extra energy in a line of affected ether electrons (photon) to the frequency of the radiation and have to do with the inertia of an electron toward angular rotation.

The following is an attempt to explain Planck's constant, on this basis:

The extra energy on an activated matrix electron would be:

$Ee = \frac{1}{2} Ie W^2$

Ie = moment of inertia of the electron to that type of spin and W = the extra rate of rotation of an activated electron.

The energy of the photon would be

$Ep = Ne \frac{1}{2} Ie W^2$

Ne = The number of activated electrons that make up the photon at any instant. The extra rate of rotation of an activated electron would be expected to increase linearly with the frequency of radiation or

$W = A f$

A is a constant and f is the frequency. Then

$Ep = Ne \frac{1}{2} Ie A^2 f^2$

or $Ep = (\frac{1}{2} Ie A^2)(Ne f) f$

but $Ep = h f$ where h = Planck's constant

then $h = (\frac{1}{2} Ie A^2)(Ne f)$

This suggests that:

1. Ie is a constant, indicating that the dimension of an aether electron does not change as its rate of rotation varies.
2. The number of electrons involved in a photon at any given instant must be proportional to the frequency of the radiation or h would not be a constant. This might be expected, since the number of electrons adjacent to a ½ wave antenna reduces linearly with the wave length and inversely with the frequency of the radiation.
3. The proposed photon energy involves integer numbers of electrons. An event can occur only if the energy required can be removed or supplied. The fewer electrons in a photon, the greater the energy difference between possible photons. This may explain the quantization of energy transfer.

My Articles Which Led to the Above Conclusions

These articles are not in chronological order. The order is based on my guess as to what the reader would prefer. Most of the articles include appropriate references to the scientific literature. The articles were written and, in some cases published, between 1985 and 2009. The reader will notice considerable repetition and, perhaps some differing conclusions. My present conclusions are those given above. I would be pleased to consider alternate conclusions. The words aether and ether refer to the same matrix. Neither refers to the chemical, which, like this book, tends to put people to sleep.

MY ARTICLES WHICH LED TO THE ABOVE CONCLUSIONS

A History of Dark Matter?

Abstract

In developing their wave equations, both Huygens and Maxwell assumed space was filled with touching material particles. Since their equations correctly predict important properties of light, their concept of a material ether were accepted as fact, until early in the twentieth century.

Several experimenters, including Sir J.J. Thomson, reported the appearance of surprisingly quantities of hydrogen gas during electrical discharge in vacuum. Clarence Skinner reported that during discharge in pure low-pressure helium, hydrogen gas was produced at the cathode and the initial rate of hydrogen production obeyed Faraday's laws of electrolysis. He obtained thousands of times more hydrogen from a silver cathode than it could have originally contained.

Recently, scientists have produced Bose-Einstein condensed (BEC) rubidium, sodium and lithium. They found that they transmit light at much lower speeds than vacuum. Could dark matter be Bose-Einstein condensed hydrogen and the medium for light transmission?

According to Linus Pauling, atomic hydrogen is paramagnetic, because it contains an unpaired electron. If Bose-Einstein condensed hydrogen is a matrix of protons and unpaired electrons, it is paramagnetic and has dielectric properties.

Such a matrix permits simple explanations for many observed phenomena including:

- The appearance of hydrogen gas during electrical discharge in vacuum.
- The values of dielectric constant and permeability of vacuum required by Maxwell's equations.
- The forces between separated magnets and
- The formation of hydrogen in space.

Note: Direct quotes from the scientific literature are between quotation marks.

Discussion
Christiaan Huygens (1678) "Treatise on Light"

Huygens[1] developed equations that predict diffraction patterns[2] based on the following assumptions:

A medium for light transfer is present in vacuum. It is made up of touching material particles. Energy is passed from particle to particle. The particles are extremely fine and easily pass through the spaces between the atoms of materials. Huygens suggested the energy is transferred much like the transfer of energy from sphere to sphere in a series of suspended balls. All the energy on one ball is transferred to an adjacent ball. The velocity of transfer depends on the properties of the balls.

"And it must be known that although the particles of the ether are not ranged thus in straight lines, as in our row of spheres, but confusedly, so that one of them touches several others. This does not hinder them from transmitting their movement and spreading it always forward."

---

1   Christiaan Huygens is a famous scientist, who was a contemporary of Isaac Newton.

2   Diffraction patterns of dark and light areas are formed when light from a single source passes through two or more tiny openings onto an otherwise dark plate. Huygens developed formulas that predict effects of the shape and positions of the holes and position of the dark plate on the designs produced.

"And this last point is demonstrated even more clearly by the celebrated experiment of Torricelli, in which the tube of glass from which the quicksilver[3] has withdrawn itself, the remaining void of air, transmits light just the same as when air was in it. For this proves that a matter different from air exists in this tube[4]. And that this matter must have penetrated the glass or the quicksilver, either one or the other, though they are both impenetrable to the air. And when, in the same experiment, one makes the vacuum after putting a little water above the quicksilver, one concludes equally that said matter passes through glass or water, or through both."

Torricelli was a contemporary of Galileo and a fellow resident of Pisa.

James Clerk Maxwell (1891)
A Treatise on Electricity and Magnetism, Volume 2

Maxwell[5] developed equations that predict various properties of electromagnetic radiation[6] using the following assumptions:

---

3   Quicksilver is mercury.
4   Sound does not travel through a vacuum. It requires a medium of particles (gas, liquid or solid). What carries light energy?
5   James Clerk Maxwell is a famous scientist who contributed to many fields of physics. He developed equations used to predict reflection, bending and absorption of electromagnetic radiation under various conditions. His wave equations predict the properties of electromagnetic frequencies that he didn't know existed.
6   Electromagnetic radiation includes radio, radar, light, x-rays and gamma rays. The main difference is the frequency of the radiation. The higher the frequency: the more energetic the radiation. Frequency is the number of times the direction of the current reverses in a second. Direct current ~ 0 cycles per second, Alternating current = 60 cycles per second, AM radio approximately 1million cycles per second, Visible light approximately one thousand trillion cycles/second.

A medium for light transfer is present in vacuum. It is made up of touching material particles. It has specific magnetic[7] and dielectric properties[8].

"In several parts of this treatise an attempt has been made to explain electromagnetic phenomena by means of mechanical action transmitted from one body to another by means of a medium occupying the space between them. The undulatory theory of light also assumes the existence of a medium."

"According to the theory of undulation, there is a material medium which fills the space between the two bodies and it is by the action of contiguous parts of this medium that the energy is passed on, from one portion to the next, til it reaches the illuminated body."

"Let us determine the conditions of the propagation of an electromagnetic disturbance through a uniform medium, which we shall suppose to be at rest, that is, to have no motion except that which may be involved in electromagnetic disturbances. Let C be the specific conductivity[9] of the medium, K its specific capacity[10] and u its magnetic 'permeability'[7].

---

7   Magnetic properties are generally attributed to unpaired electrons in a material. Electrons are extremely tiny negatively charged particles. They are present in all materials. If an electron is paired with another electron, its magnetic properties are canceled. Permanent magnets contain unpaired electrons that are not randomly oriented. Paramagnetic materials contain randomly oriented, unpaired electrons. In the vicinity of a permanent magnet, some of the unpaired electrons in a paramagnetic material become non-randomly oriented. The paramagnetic material behaves like a permanent magnet, until the permanent magnet is removed.

8   All materials contain positive and negative particles. When a material is placed between oppositely charged metal plates, positive particles tend to move toward the negative plate. Negative particles tend to move toward the positive plate. This is equivalent to the storage of electrical energy. The relative dielectric constant of a material is the ratio of the electrical energy stored by the material to that stored by vacuum.

9   Specific conductivity refers to the speed of energy transmission through a medium.

10   Specific capacity is related to dielectric properties. Maxwell's wave equations require that vacuum have definite magnetic and dielectric properties. Can a void have such properties? The relative dielectric constants of materials reduce dramatically as the frequency of electromagnetic radiation increases. This is attributed to the inertia of some of its particles to movement. At some frequencies, some particles are unable to move fast enough and no longer

C. A. Skinner (1905), J.J. Thomson (1914) and G. Winchester (1914)

Clarence Skinner performed electrical discharge[11] experiments in low-pressure helium. He reported hydrogen gas was produced at the cathode and the initial rate of production followed Faraday's laws of electrolysis[12]. He obtained thousands of times more hydrogen from a silver cathode than it could have originally contained and stated, "It shows no sign of having its supply of hydrogen reduced in the least". He reported that most cathodes tarnish during electrical discharge in pure helium. The following is the introduction to his excellent article.

"While making an experimental study of the cathode fall[13] of various metals in helium it was observed that no matter how carefully the gas was purified the hydrogen radiation, tested spectroscopically[14] persistently appeared in the cathode glow. Simultaneously with this appearance there was also a continuous increase in the gas pressure with time of discharge. This change in gas pressure was remarkable because of its being much greater than that which had been observed under the same conditions with nitrogen, oxygen or hydrogen. Now the variation in the cathode fall, with current density and with gas pressure, in helium, was found to be so like that obtained earlier with

---

contribute to dielectric properties. The dielectric constant of vacuum does not change with frequency. If it contains particles, they must have an extremely low moment of inertia. This suggests that they must be extremely small in size and have very low masses. Electrons have just such properties.

11   Discharge tubes are similar to the tubes employed in neon signs. Skinner's tubes were straight, with a metal electrode at each end. The negative electrode (cathode) was at one end and the positive electrode (anode) was at the other end. The tube was evacuated and pure helium (or other gas to be tested) was introduced into the tube to a pressure of about three thousandth of atmospheric. The voltage across the tube was increased until glows appeared. The voltage was gradually reduced to keep the current passing through the tube at 0.003 amperes.

12   Michael Faraday contributed greatly to many fields of physics. He developed equations that predict the amount of a product that will be produced when a constant electric current is passed through a material.

13   Cathode fall is the difference in voltage between the cathode and a position, in the tube, between the cathode and the anode.

14   Each gas produces a specific spectrum when subjected to electrical discharge. The presence of hydrogen gas is easily detected by its spectra.

hydrogen that it appeared necessary to maintain the helium free of the latter in order to make sure that the hydrogen present was not the factor causing this similarity in the results. Futile endeavors to attain this condition led to the present investigation, which locates the source of hydrogen in the cathode, shows that the quantity of hydrogen evolved by a fresh cathode obeys Faraday's law for electrolysis, and that a fresh anode absorbs hydrogen according to the same law."

(In this book I present a different mechanism for the apparent removal of hydrogen. See "My Correspondence with Linus Pauling, Part2")

"Altogether about two cubic centimeters of gas[15] have been given off by this silver disk, which is 15 mm. in diameter and about one mm. thick. It shows no sign of having its supply of hydrogen reduced in the least."

In experiments leading to the development of the mass spectrograph, Sir J.J. Thomson[16] was unable get his discharge tubes free of hydrogen.

"I would like to direct attention to the analogy between the effect just described and an everyday experience with discharge tubes. I mean the difficulty of getting these tubes free from hydrogen when the test is made by a sensitive method like that of positive rays[17]. Though you may heat the glass tube to the melting point, may dry the gases by liquid air or cooled charcoal and free gases you let into the tube as carefully as you will from hydrogen, you will get hydrogen lines by the positive ray method, even when the bulb has been running several hours a day for nearly a year."

Since the gases tested by Thomson were subjected to electrical discharge prior to test, he may have produced hydrogen by the same mechanism as Skinner. If the medium proposed by Maxwell is a matrix of protons and unpaired electrons, atomic hydrogen might be

---

15   This is over 1000 times as much hydrogen than the silver cathode could have originally contained.

16   Sir J.J. Thomson was awarded the Nobel Prize for his work. He was knighted for discovering the electron and determining its mass along with other important contributions to science.

17   Positive rays are positive ions produced in discharge tubes. By passing such ions through electric and magnetic fields, Thomson was able to measure the mass and relative concentration of each ion. His work led to the development of the mass spectrometer.

produced from the medium by electrolysis. If so, atomic hydrogen would be produced at a fresh cathode at the rate predicted by Faraday's Laws. Atomic hydrogen is extremely reactive and would be expected to tarnish metal cathodes as noted by Skinner.

George Winchester performed electrical discharge experiments at pressures as low as one millionth of a millimeter[18]. He employed aluminum electrodes, with a minimum gap and about 100,000 DC volts. His paper includes graphs showing the rate of pressure increase under various conditions. He obtained hydrogen, helium and neon. Eventually the production of helium and neon ceased.

"The case of hydrogen is different; I have sparked tubes until the electrodes were entirely wasted away and this gas can be obtained as long as any metal remains."

Preliminary Conclusions

1. The medium for light transfer has definite magnetic and dielectric properties. Dielectric properties suggest that the medium contains positive and negative particles. Magnetic properties suggest that the negative particles might be unpaired electrons.
2. Hydrogen gas appears to be formed by electrolysis of the medium. This suggests that the positive particles may be protons.

Linus Pauling (1945) "Nature of the Chemical Bond"

"The most stable orbit in every atom is the 1s orbit of the K shell. In the normal hydrogen atom this is occupied by one electron, the spin magnetic moment of which makes monatomic hydrogen gas paramagnetic."

"It is customary to refer to electrons with opposed spins as paired, whether they occupy the same orbit in one atom or are involved in the formation of a bond."

---

18   One millionth of a millimeter of mercury pressure is 0.00076 th. of atmospheric pressure.

Second Conclusions

1. The magnetic properties of vacuum suggest that vacuum contains unpaired electrons and is, therefore, paramagnetic. This leads to a simple explanation for the forces between two separated permanent magnets.
2. The medium may be a matrix of protons and unpaired electrons, much as in molten common salt, where no sodium ion touches another sodium ion and no chloride touches another chloride ion.

Weidner and Sells (1969)
Elementary Modern Physics

"We shall see that, apart from the tremendous difference in their relative sizes, 10^{-10} m. [19] for atoms but less than 10^{-14} m. for nuclei, nuclear structure is different from atomic structure in several significant respects." (Note the factor of 10,000).

Every nucleus includes at least one proton. The size of the proton must be less than 10^{-14} m. The classical radius of the electron is about 10^{-15} m. The space between nuclei of materials is a wide-open gate to a matrix of touching protons and electrons, as proposed by Huygens. Glass transmits light. Likely, it contains the medium for light transmission. The medium may permeate all materials. Whether a material transfers light depends on its dielectric and magnetic properties, at the frequency concerned. (Maxwell)

Bose-Einstein Condensation (BEC)
Silvera and Walreven (1982)

The following quotes are from an article in the January, 1982 issue of Scientific American:

"The statistical theory that describes atoms was first studied by the Indian physicist S.N. Bose and is called Bose-Einstein statistics. The phenomenon predicted by Einstein is a mathematical consequence of Bose Statistics, but it was so contrary to the intuition

---

[19] $10^{-1} = 0.1$, $10^{-2} = 0.01$, $10^{-3} = 0.001$, etc. The symbol for meters is m.

of physicists in the 1920's that it was regarded as a mathematical oddity that would never be found in a real system. It is now thought the phenomenon is observable in the laboratory. It is called Bose-Einstein condensation."

"The critical temperature for the condensate is proportional to the density raised to the 2/3 power"

Based on the masses and sizes of protons and electrons Bose-Einstein condensed hydrogen would be stable at extremely high temperatures.

"Liquid helium 4 at or below 2.18 degrees[20] is therefore called a superfluid. If it is set flowing in a tube closed on itself the liquid continues to flow without friction, never coming to a stop as a normal fluid would. It flows into the smallest passages of its containing vessel and has the remarkable ability to flow through a densely packed powder as if the barrier was not present. A vessel with microscopic holes that would be impenetrable to a normal fluid can be a leaky sieve to a superfluid. Such a vessel is said to have a superleak."

Daniel Kleppner[21] and Thomas Greytak

The following quote is from the December, 2000 issue of Scientific American:

"When his former students were making their spectacular condensates of rubidium, sodium and lithium (alkali metals)[22], Kleppner was battling his career-long atom of choice; hydrogen. He has been studying hydrogen since he was a graduate student and postdoc at Harvard University in the late 1950s."

Their experiments toward producing Bose-Einstein condensed hydrogen appeared to be unsuccessful until they employed spin-polarized hydrogen[23]. Perhaps, they had produced Bose-Einstein

---

20 Temperatures in this speech are in degrees absolute. Zero degrees absolute is the lowest possible temperature.

21 Daniel Kleppner was awarded the Nobel Prize in physics for this work.

22 Hydrogen, lithium, sodium and rubidium have similar structures. They are in the first column of the periodic table. When they ionize, they tend to form a singly positive charged ions and electrons.

23 Spin polarized indicates that all the hydrogen atoms are in the same alignment.

condensed hydrogen earlier, but couldn't detect it. It is difficult to detect water you have produced in a summer lake. Ice is much easier to detect.

Lene Vestegaard Hau

The July 2001 issue of Scientific American includes an article by Lene Vestegaard Hau. It describes experiments her group performed at the Rowland Institute in Cambridge, MA. They passed laser beams into Bose-Einstein condensed sodium and found that it transferred light at a much lower speed than vacuum. They were able to stop light transmission and, then restart it, at will, using appropriate laser beams.

If Bose-Einstein condensed sodium transfers light, one might expect Bose-Einstein condensed hydrogen to transfer light at a much greater velocity[24].

Final Conclusions:

Assuming knowable space is permeated with a concentrated matrix of protons and electrons (possibly Bose-Einstein condensed hydrogen) permits simple explanations for many observed phenomena including:

1. The stability of galaxies[25].
2. The formation of hydrogen in low-pressure gasses.
3. The values of dielectric constant and permeability of vacuum required by Maxwell's equations.
4. The forces between separated permanent magnets.
5. The formation of hydrogen in space and

---

24 Bose-Einstein condensed sodium is expected to be much denser than Bose-Einstein condensed hydrogen. Its dielectric constant should be much greater than Bose-Einstein condensed hydrogen. Light would be expected to pass through the sodium condensate at a much slower rate.

25 In order to explain the stability of galaxies, scientist have introduced the concept of dark matter. Could their dark matter be Bose-Einstein condensed hydrogen and the ether of classical physics?

6. If transmission of light is less than 100.0000000 … %
efficient, the "red shift" might be caused by the lost
energy. (E = Planck's constant X frequency[26],[27]).

Paul E. Rowe 02649rowepaul@comcast.net,
May 22, 2007

I suspect Planck's constant is based on the inertia of the electron
to the rotation required for light transmission.

---

26   The spectra of light from distant parts of the universe have shifted toward
the red. For this reason, scientists believe the universe is expanding. If any
energy is lost in traveling through the matrix, a similar red shift might be
expected.
27   Light is believed to pass through space as packets or photons. E is the
energy of the photon. Planck's constant is a universal constant employed to
make this and many other equations useful.

My Conversations with
Einstein (in Dreams)

By Paul E. Rowe 02649rowepaul@comcast.net
May 29, 2008

Introduction:

In 2004, I wrote a play, "The Fall and Rise of the House of Cards".
It included conversations I had (in dreams) with various deceased
scientists. The play was so long and so dull that no one could read
more than six pages and stay awake.

The play suggests the knowable universe is permeated with a
concentrated matrix of protons and unpaired electrons, possibly
Bose-Einstein condensed hydrogen. Could this be "Dark Matter"
and/or the ether of classical physics?

This paper includes the beginning of Act II and my conversations
with Albert Einstein.

I hope someone will be able to read this part of the play. If anyone
has trouble sleeping, he or she should try to read the rest of play. It is
included later in this book.

Act II, Scene I

Curtain rises to show Paul sitting at the desk, writing in his notebook.
He turns to the audience.

Paul: Welcome back. Let's have a show of hands. How many of you are completely confused by this play? Sorry about that! Now, is there anyone here who follows everything, understands completely, and believes every word? Well, are any of you actors impressed? Ok. Perhaps a short summary is in order:

1. I obtained much more gas than was theoretically possible by detonating explosives containing aluminum flake in vacuum.
2. A literature search revealed that many experimenters, some quite famous, reported obtaining surprising amounts of hydrogen gas in their experiments.
3. I produced hydrogen gas by igniting mixtures of cupric oxide and aluminum powder in vacuum.
4. Like Skinner, I produced hydrogen gas during electrical discharge in low, pressure helium. I also produced hydrogen in a fairly good vacuum.
5. I produced hydrogen gas by other techniques, including placing a glass tube containing a fairly good vacuum near an operating spark coil.

This has led me to believe that vacuum is not a void. It contains something that can be converted into hydrogen, under the proper conditions. My best guess is that the knowable universe is permeated with a matrix of protons and electrons. This may be aether, the light-carrying medium, that scientist accepted as fact, in the late 19 th. and early 20 th. centuries. If this is the case, why isn't it obvious to everyone? How can I move my hand through the matrix with such little effort? I performed another literature search in an effort to answer these questions.

Later in Act II

Bohr: That is the response of a true scientist. By the way, you haven't explained gravity yet. (Albert Einstein enters at the back of the stage and stacks cards above the base that Bohr had assembled.)

Paul: I've been trying to build the proper background. I'm happy with the idea that magnetic forces are transferred through the electrons of the aether. I am not as happy with the idea that gravity is transferred through the protons of the ether. However, if one accepts that the neutron is a combination of a proton and an electron, the weight of any material is very close to the weight of the protons it contains. If one assumes that there is an attractive force between protons that remains after electrical forces are cancelled by electrons, one can come very close to explaining gravity.

Bohr: You are not happy with that description of gravity?

Paul: Not very. It needs more work.

Bohr: I see that you have a distinguished visitor. If anyone can pick apart your suggestions, it is Professor Einstein. (To Einstein.) Hello Albert. I assume you are here to take your turn. (To Paul.) Goodbye Paul. I enjoyed your monologue. If you eventually make it to heaven, it will be my turn to do the talking. (As Bohr leaves, he pauses to admire the enlarged house of cards.)

Paul: I'm sorry. Sometimes I get carried away. I look forward to your monologue, but I'm in no real hurry. (To Einstein) It is a great honor to meet you Professor Einstein. I won't introduce you because your fame has preceded you.

Einstein: It is a pleasure to meet you Paul. You needn't be so formal. Please call me Albert. .

Paul: Well, Albert, I hope you don't mind if I call you a genius. I have nothing but respect for your mathematics, but I would like to discuss some of your interpretations. After all, some of them are close to 100 years old. I have the advantage of the results of many more years of research by brilliant scientists.

Einstein: To prove that I have an open mind on these matters, I will quote from a letter I wrote to a fellow professor in March of 1949:

> "You can imagine that I look back on my life's work with calm satisfaction. But from nearby it looks quite different. There is not a single concept of which I am convinced that it will stand firm, and feel uncertain whether I am in general on the right track."

45

Don't worry about my feelings. Your only goal should be to improve mankind's understanding of its universe.

Paul: I expected you to feel that way. Before we begin, I'd like to ask you a question. Do all great scientists end up in heaven?

Einstein: Of course not. Why do you ask?

Paul: I've conversed with many scientists in this play. I assume that they are all in heaven.

Einstein: Not necessarily. If someone from the other place wishes to be involved in a dream, he may petition St. Peter. If permission is granted, he may enter dreams. There is a problem, however. Their internal thermostats are adapted toward a different climate and they are extremely uncomfortable when away from hell. According to the "Guinness Book of Records" no one from there has been able to last more than 5 minutes in a dream.

Paul: It sounds like death is more complicated than I thought. Let me summarize my position. I am convinced that hydrogen gas has been created from vacuum. This leads me to suspect that vacuum contains something from which hydrogen can be produced. Other observations led me to suspect that vacuum contains a concentrated matrix of protons and electrons. Such a matrix (the aether?), agrees with the ideas of Huygens and Maxwell on the nature of light. The presence of unpaired electrons would cause the matrix be paramagnetic. This permits simple explanations for the forces between magnets separated by vacuum and accounts for the permeability of vacuum required by Maxwell's wave equations. I do not believe that you explained this effect in your papers. I hope you do not object to my taking advantage of your work in making my case.

Einstein: I would be thrilled, if you present a reasonable case.

Paul: I'll start by referring to Bose-Einstein condensation. (To the audience) Professor Einstein and Satyendra Nath Bose of India collaborated in developing Bose-Einstein statistics. This led them to conclude that, under the proper conditions, materials may form condensates of particles in their lowest energy states. They may remain in that state at temperatures above absolute zero. Liquid helium below 2 degrees Kelvin has strange properties and many scientists believe that it is Bose-Einstein condensed. It flows as if it were a zero viscosity liquid. For example, once started in motion through a granular material, it will continue to flow at the same speed indefinitely. This may remind you of Newton's conclusion that a body in motion will continue in motion, until it is affected by an external force. In other words, your hand can move through a Bose-Einstein condensate without restriction. Dr. Einstein explained this approximately as follows:

A Bose-Einstein condensate is at its lowest possible energy state at a temperature above absolute zero. It cannot absorb energy and remain in this state. He developed mathematical formulas that predict the highest temperature at which a material can remain Bose-Einstein condensed. Based on the weights of protons and electrons and their separation, my proposed matrix should remain Bose-Einstein condensed at extremely high temperatures. Such a condensate cannot absorb energy by the usual mechanisms.

The proposed matrix may be considered a polymer of hydrogen atoms that, though made up of charged particles, is neutral, in the sense that chemical salts, which are made up of oppositely charged ions are neutral. It seems (to me) to have the properties scientists require for dark matter.

Einstein: I admit to being surprised by the conclusions to which our formulae led us and doubted that anything would come of it.

Paul: Recently, scientists have produced Bose-Einstein condensates of rubidium, potassium, lithium and sodium. These condensates transfer light, but at much lower speeds than vacuum. There was surprising difficulty in producing a similar condensate from hydrogen. More recently, a group led by Professor Daniel Kleppner, at MIT, formed Bose-Einstein condensed hydrogen but he had to have all the protons aligned in the same orientation. I suspect that he and other scientists had previously produced the condensate of hydrogen, but couldn't detect it, because they produced aether, which fills space. If you produce a little water in a lake, you would have a hard time detecting it. If you formed ice in the summer, you could easily detect it. The condensate with aligned protons may have been different enough to be detected.

Einstein: And people accuse me of having strange ideas. Let's see you explain the results of the Michelson-Morley interferometer experiments.

Paul: I'm surprised you mentioned that. (To audience.) In the late nineteenth century the presence of an aether that carried light was accepted as a scientific fact. It was assumed that the earth, in its orbit around the sun, was moving rapidly through the aether. If this were the case, light moving through the aether should travel at a fixed velocity relative to the aether, but at different velocities relative to the earth, depending on the direction the light traveled, relative to the earth. Their measurements seemed to show that the direction light traveled, relative to the earth, did not affect its velocity, relative to the earth. This was an important event in discrediting the aether concept. In spite of his results, Michelson continued to accept the presence of a material aether. In the 1920s and thirties, Sir Oliver Lodge continued to champion the ether concept. He pointed out that the results of the Michelson-Morley experiments are just as would be expected, if the portion of the aether in the vicinity of the earth moved along with the surface of the earth. The matrix of protons and electrons that I have proposed would have mass and would be expected to tend to move with the surface of the earth as do air and water.

Einstein: Why are you surprised that I mentioned Michelson-Morley?

Paul: Page 96 of the book, "Einstein" by Ronald W. Clark refers to statements you made which suggest that the Michelson experiments had little influence on you in developing relativity and that the experimental evidence which had influenced you most were the observation of stellar aberration and Fizeau's measurements of the speed of light in moving water.

Einstein: Actually, I brought up the Michelson-Morley experiments because they are still referred to in your time. Do you have any comment on stellar aberration or on Fizeau's experiments.

Paul: Yes. It seems to me they both tend to corroborate Lodge's point of view.

Einstein: I would like to hear your explanations.

Paul: I'll discuss stellar aberration first. (To audience) When astronomers on earth try to fix the position of stars in space they correct their measurement for known effects like bending of light as it enters the earth atmosphere. In spite of these corrections, the position of any star seems to change slightly as the earth rotates around the

48

sun. In the northern hemisphere the North Star appears to move in a small circle. The apparent position depends on the date of the year on earth. No one believes the North Star's actual position is affected by the date on the earth. This effect is called stellar aberration.

(To Einstein.) If the portion of the aether in the vicinity of the earth moves with the earth as it orbits around the sun, at about 26 miles per second, light from a star bends as it approaches the earth. A similar effect occurs when a sound wave encounters wind. In the case of starlight, the direction of bending, relative to the star, reverses every six months. In the case of the North Star, one would expect the deviation to be circular. The observed position should be the same on the same date each year.

Einstein: Don't you think I considered that possibility?

Paul: I don't know. Did you?

Einstein: Possibly not. How do you explain Fizeau's results?

Paul: Let me discuss this with the audience. (To audience) Armand Fizeau was a French physicist. In 1849, he determined the first reasonably accurate velocity of light. He also showed that, when light traveled through moving water, its velocity was greater, relative to the earth, when it moved in the same direction as the water, than when it moved in the opposite direction. This suggests, to me, that the aether, which carries light, tends to move with the molecules of water. I consider this as corroboration of Lodge's contention that the aether near the earth moves with the earth.

(A man in a state of anxiety runs onto the stage and approaches Einstein)

Man: (In French.) You see. This is what I have been telling you. My experiments were carried out to prove the existence of the aether. Somehow, you used them to discredit the aether concept.

Einstein: (To Paul.) Do you understand French?

Paul: Very little.

Einstein: Then let me talk with Armand and then translate for you.

Paul: Thank you. I would appreciate that.

Einstein: (To Fizeau. In French.) I'm sorry that you find my using your work, bothersome. I assure you that was not my intention. You must appreciate that anything in the scientific literature is fair game.

Anyone can interpret it as he wishes. I am sorry that you died 10 years before I published my first relativity paper. I would have been happy to discuss these matters with you.

Fizeau: (In French.) I know that you are correct in this matter, at least. Your misinterpretation of my work has been bothering me for some time. I could not miss this opportunity. I am not feeling well. I have to leave. (He starts to leave. While leaving, mutters in French.) Stupid Kraut.

Paul: Wow! That was strange. What did he want?

Einstein: He wanted to tell me off for misinterpreting the meaning of his experiments. He felt that they went along with Lodge's ideas and, of course, yours. I tried to be nice and he cooled down somewhat.

Paul: What was the reference to Kraut?

Einstein: He called me a stupid Kraut, as he left. He wasn't feeling well. I think it bothered him that he couldn't stay for over five minutes and get in the "Guinness Book of Records".

Paul: I wish he could have stayed longer. It is nice to find a scientist who agrees with one of my ideas.

Einstein: Perhaps you will meet him again someday.

Paul: If I am interpreting correctly, I hope not.

Einstein: You seem to understand. Is there anything else you wish to mention before I leave you?

Paul: I'd like to consider your ideas on the effect of velocity on mass and your twin paradox. I have much more to discuss with you, but I don't want to impose.

Einstein: I am enjoying our conversation but I must leave soon.

Paul: I'll try to be brief. As an airplane approaches the sound barrier, more and more energy is required to impart a given acceleration. The mass of the airplane is believed to remain constant. A greater force is required because the air through which the airplane is traveling increases its resistance to increased velocity. The medium for light may produce similar effects, as a rocket ship approaches the "light barrier". This is possible if the energy increase is in the material and not transferred to the aether. Huygens believed that matter is an open mesh through which the aether easily passes. However, as an object approaches the speed of light, aether particles may have increasing difficulty moving around the nuclei of its atoms. If so, the energy required for imparting a given acceleration increases with increasing velocity.

If one defines a unit of mass based on a standard that is at rest, relative to the aether, in its vicinity, one expects greater energy to be required to accelerate an object as the object approaches the speed of light. By this definition of mass, the force required to accelerate an object increases as the object approaches the speed of light, its mass does not change as you have proposed.

Einstein: How about my twin paradox? (To the audience) My mathematics led me to conclude the following:

If one of a pair of twins flew away from the earth at close to the speed of light and later returned at a similar velocity. He would appear to be and actually would be younger than his twin. In other words, time slows as velocity increases.

Paul: If materials that make up a rocket ship are porous to the ether, a pendulum clock in a rocket ship moves through the ether at a similar rate as the rocket ship, itself. As the clock approached the speed of light, the period of the pendulum would decrease and time, as measured by the clock, would decrease. The same effect would occur with modern time devices that are based on the periods of certain bond vibrations.

In the twin paradox, one twin remains on earth while the other twin rockets away at close to the speed of light, turns around and then returns at a similar speed. The speeding twin's clock would run slower in both directions of flight, since in each case, it is moving rapidly through the aether. The earthbound twins clock is comparatively still, relatively to the aether, in its vicinity, and would indicate the passage of more time. At the speed of light the pendulum and the vibration might stop.

I prefer a definition of time based on a clock that is stationary, relative to the aether, in its vicinity. In order to determine the correct time, a factor, based on the speed at which a clock is moving through the aether (or, if you prefer, the aether is moving through the clock) would be applied to the time it registers. Time would, then, be more nearly universal and independent of velocity.

If I am correct, the vibrations of the chemical bonds in the speeding twin's body would be slowed dramatically. This would slow all the chemical reactions in his body. If he survives, he may well appear younger than the stationary twin.

Einstein: You covered those subjects in a very short time. You can mention one more topic, if you hurry.

Paul: Great. Let me give you my guess as to the nature of the photon.

Einstein: Go to it.

Paul: When a DC current passes through a straight wire, a compass needle near the wire points perpendicular to the wire. It remains so oriented until the current is stopped. This suggests that moving electrons on the wire cause aether electrons in the vicinity of the wire to orient. When the current is reversed, the needle points in the opposite direction. At extremely low frequencies, the orientation of the needle changes with frequency. At higher frequencies, the needle of the compass cannot reorient fast enough, because of its inertia . Aether electrons have very little inertia and continue to orient with changing current direction, even at extremely high frequencies. (Could Planck's constant be related to the inertia of the electron to the type of rotation required?) Let us consider a half-wave dipole antenna operating at a FM frequency. I understand that the current moves along the surface of such an antenna and at the speed of light. According to Huygens' principle, each orienting ether electron passes all of its energy to a neighboring ether electron, etc. This effect moves away from the antenna at the speed of light.

As the direction of the current changes, a line of energy starts forming at one end of the antenna. Just as the direction of the current changes again, this line of energy leaves the other end of the antenna. The result is a line of energy moving through the ether at the speed of light (a photon?). This line is in a plane of the antenna and at an angle of 45 degrees to the antenna.

When the direction of the current changes again, a similar line of energy (photon?) is produced. This photon is also in a plane of the antenna but at an angle of 135 degrees to the antenna. Its aether electrons are orienting oppositely. What we consider one wave in the antenna may produce a pair of photons. Each photon might be considered a mirror image of the previous photon.

At any instant, an integer number of active electrons are involved in a photon (as I pointed out earlier there are no fractional electrons). For this reason, only specific amounts of energy can be carried as photons. As I suggested to Professor Bohr, this may be the basis for quantum mechanics.

Einstein: You know Paul, I am considered one of the most brilliant scientists ever and I can't keep up with the pace in which you have presented your ideas. Do you think anyone in the audience understands you.

Paul: No. If they are interested, I will give them a good deal on a copy of the play. I have to read such proposals a few times and then, sleep, before I understand them. If I made it too easy, there would be no incentive for them to buy a copy of the play. As a bonus, they may have some interesting dreams. Perhaps, the following summary will help:

Assuming that the knowable universe is permeated with a zero viscosity medium (possibly, Bose-Einstein condensed hydrogen) leads to simple explanations for many, otherwise, difficult to understand phenomena. Such explanations may permit intelligent laymen to achieve a clear understanding of theoretical physics. They may also lead scientists to more efficiently direct their efforts. Acceptance of such an aether would require us to reconsider concepts developed based on the absence of an aether.

Einstein: I hope that helps. I will ponder your ideas and hope to discuss these matters with you in the distant future. By the way, I was never able to successfully combine the ideas of relativity and quantum mechanics. Can you combine your ideas and quantum mechanics.

Paul: I purposely avoided mathematics in this play. I suggested, in a paper I wrote in 1998, that Planck's constant may be related to the inertia of an aether electron toward the angular rotation required for transmitting photons through the aether.

Paul: Meeting you in the distant future sounds great to me. See you then. Thanks for listening. You've been very kind.

Einstein: Goodbye, Paul. (He walks off stage. Paul gets in the cot and pulls up the blankets and starts snoring. There is a rumbling sound and things on stage start to shake.)

Einstein: (Shouting.) Neils, come here. I need you. We are having an earthquake. (Professor Bohr runs onto the stage and helps Einstein protect the house of cards.)

(The curtain closes.)

Einstein: (Shouting.) Gott in Himmel! (The sound of falling cards is heard throughout the theater.)

Act II, Scene II

The curtain opens to show Paul sleeping and snoring. The lights are normal, indicating this is not a dream sequence. The cards are a mess.

Kate: (Off stage.) Paul. Are you down there? (Paul stirs and opens his eyes.)

Paul: Yes. I'm here.

Kate: Diner is ready.

Paul: Great I'll be right up. I had the craziest dream. I'll tell you about it.

Kate: Was it about the aether.

Paul: (Getting up and walking back stage,) Yes.

Kate: Then don't bother. I thought you were going to wash windows today.

Paul: Oh! That's right. I forgot. (Looks at the messed up cards. Scratches his head. Turns to the audience and shrugs his shoulders.) (To Kate) I'll wash them tomorrow, right after I pick up the cellar. (Walks offstage.)

The Curtain Falls

Act II, Scene III

(The same stage setting on the next evening. Paul enters and speaks to the audience.)

Paul: Boy, am I tired? It is been a beautiful day, so Kate suggested that I wash the windows before picking up the cellar. It is always easier to go along with her suggestions than to argue. Please excuse me. I have earned a little nap before straightening out these cards. (Paul lies on the cot and, after a few seconds, starts to snore. The lights become bluish. Einstein enters and shakes Paul to wake him. (Paul stirs, stretches and yawns. He looks up and sees Einstein).

Paul: Good evening Albert. I thought you left.

Einstein: I did but my conscience bothered me. I decided to return and help you pick up the cards.

Paul: Thank you. I'll appreciate any help. Before we start, I would like to discuss a dream I just had.

Einstein: That would be fine, but I can't stay for long.

Paul: I'll try to be quick. I understand that an important basis of your special theory of relativity was Maxwell's field equations.

Einstein: That's true. I may not have conceived of special relativity without them.

Paul: In my conversation with Maxwell, I quoted from one of his books. In developing these equations, he assumed that light was transferred through a medium of touching particles.

Einstein: So?

Paul: Well, if you employed his equations in developing your theories weren't you using the same assumptions? I've heard that you felt your equations described physical phenomena without assuming a physical medium in vacuum. You have been quoted as saying, "If a thing can't be observed, why should it be necessary to assume its existence?"

Einstein: You are making me wonder if I should have returned to help you. I wish I were allowed to answer your question. As others have told you, we cannot reveal anything that we have learned after our deaths. I will try to answer questions that don't fall into this category. Where did you find that quote?

Paul: It is on page 315 of "An Outline of Atomic Physics", by Oswald H. Blackwood et al., second edition, seventh printing, published in 1946. It claimed that the quote was from 1905. The following quote is from your book with Leopold Infeld, "The Evolution of Physics", copyright, 1938, p. 152.

"In Maxwell's theory there are no material actors."

The next quote starts on page 159 of the same book:

"Our only way out seems to be to take for granted the fact that space has the physical property of transmitting electromagnetic waves, and not to bother too much about the meaning of this statement. We may still use the word ether, but only to express some physical property of space. This word ether has changed its meaning many times in the development of science. At the moment it no longer stands for a medium built up of particles. Its story, by no means finished, is continued by the relativity theory."

If space is void, can it have physical properties for transmitting light, magnetism and gravity? Can it have different properties in the vicinity of the sun than some distance from the sun? I suspect your prediction of the properties of Bose-Einstein condensates is the most important factor leading to the acceptance of a particle-based ether.

Einstein: Thank you for that. I must admit that the last quote seemed to ramble a bit.

Paul: You probably won't be able to respond to this, but let me talk to the audience. Albert predicted that light passing close to a large mass is bent toward the mass. Measurements made during a solar eclipse have been interpreted to show that starlight passing near the surface of the sun is deviated toward the sun. More recent measurements indicate that light passing near a galaxy is deviated toward the galaxy. According to Maxwell's equations, this suggests that vacuum in the vicinity of masses has a higher dielectric constant and/or permeability than vacuum in free space. If there is a medium for the transfer of light, it must be denser in the vicinity of masses and the closer it is to a mass the greater its density. This is also true of gravitational attraction. Could the medium for light also be the medium for gravity? If space is void, how can it have different properties at various distances from masses?

Einstein: You are right Paul. I may not answer those questions.

Paul: It seems, to me, that present day theoreticians are going to ridiculous extremes to retain their present beliefs concerning relativity and quantum mechanics. They are proposing the need of ten or more dimensions and weird particles that exist for nanoseconds as necessary to explain important phenomena. This reminds me of the long delay in the acceptance of Copernicus' ideas of the sun-centered solar system, due to so-called, experts using complicated mathematics. Only an elite few appeared to understand the earth centered universe. When the Copernicus sun-centered solar system was finally accepted, intelligent laymen had no difficulty with his concept.

Einstein: Are you saying that I am responsible for delaying the progress of science?

Paul: Of course not. Science has made great strides since you published your relativity theories. I couldn't have conceived how materials move effortlessly through my proposed medium without knowing about Bose-Einstein condensation. In your lifetime, you pointed out that your theories would probably not stand the tests of time. If my ideas are accepted, they will be disproved or, at least, altered in a few years. Your work has dominated science for a century.

Einstein: Thank you Paul I appreciate your comments. I'm sorry, I can't help you with the cards, but I must leave now. (Einstein starts leaving.).

Paul: That's all right, I'd much rather have had this conversation, than your help with the cards. Anyway, I've decided to build my own house of cards, but I doubt that it will be a century before the next earthquake. Thanks again for offering to help. I look forward to meeting you in the future. Before you go, I'd like to make a prediction based the density of the medium decreasing with distance from masses.

Einstein: Go right ahead.

Paul: If the medium is denser the nearer it is to a large mass, the speed of light is less near the earth than in the space farther from the earth. It is less in the solar system than it is in the space between the stars. It is less in galaxies than in the space between galaxies than it is in galaxies. Since scientists assume the speed of light in deep space is the same as that in the vicinity of the earth, the distance to other galaxies may be considerably greater than they have calculated.

Einstein: That is very interesting. I had great difficulty in getting my ideas accepted. I would have had much less of a problem, if I could have pointed out some immediate practical application.

Paul: You probably know that there is great concern today about the depletion of the earth's fuels and the deleterious side effects from our present sources of energy. All of the techniques I have discussed for converting the medium into hydrogen require the input of large amounts of energy to produce comparatively little hydrogen. Conversion of the hydrogen in water into the medium, would release considerable energy. The byproducts would be oxygen and the medium. If one could control the process, there might be no deleterious effects.

Einstein: Have you tried to accomplish this?

Paul: Yes, and I have had some interesting, but not dramatic results. I haven't been able to convince myself about the source of the energy. I certainly couldn't convince the scientific community. The source of the energy from lightning remains a mystery. I wonder if it could be the conversion of hydrogen in moist air into the medium, during electrical discharge.

Einstein: That is just the kind of thing I had in mind. I have time for one more prediction.

Paul: I'll make it quick. The electron that activates the face of the cathode ray tube may not be the electron that left the cathode. The effect may be passed from aether electron to aether electron until an electron at the face of the cathode activates a molecule on the inner face of the tube. This is similar to the mechanism Huygens proposed for light. The rate of energy transfer increases with the voltage across the tube. The electron microscope works because, on a statistical average, this is a wave phenomenon.

Einstein: Goodbye Paul. I'm off to paradise. I hope to see you there, in due time.

Paul: Auf wieder sehen Albert. I hope so, too. (As Einstein leaves, Paul walks to the back of the stage and starts building his house of cards. He sings)

Paul: I'll build a stairway to paradise, with a new step every day.

The curtain falls.
Curtain Call (If Required)
Curtain parts. Paul walks on Stage and faces Audience.

Paul: I thank you, especially anyone who is not a relative. Oh! By the way, I forgot to ask.

I didn't snore. Did I?

Curtain falls

(The recording of, "I'll Build a Stairway to Paradise", from the movie, "An American in Paris" is heard throughout the theater. It continues to play as the audience leaves the theater.)

The End

Play written August, 2004
Edited June 12, 2008
Paul E. Rowe, 71 West Way, Mashpee, MA 02649,
02649rowepaul@comcast.net

My Correspondence
with Linus Pauling

Part 1

This is a copy of the first letter I sent to Linus Pauling and his response.

34 Summit Avenue
Sharon, MA 02067
February 22, 1985

Professor Linus Pauling
Linus Pauling Institute of Science and Medicine
440 Page Mill Road
Palo Alto, CA 94306

Dear Professor Pauling;

Please find a paper which I sent to several physicists, without response. I am trained as a chemist, so I decided to send it to you. If I am to be ignored, it might as well be by the best. The following is a quick summary. I hope it will convince you to read the paper.

1. When detonating explosives containing aluminum powder, I consistently obtained much more gas than I could account for.

2. A literature search revealed that similar results occurred when tungsten wires were atomized by capacitor discharge.

3. Many experimenters reported the unexpected appearance of surprisingly large amounts of hydrogen gas during electrical discharge in low pressure gasses.

4. Clarence Skinner reported results which he felt (and I feel) show that hydrogen was produced at the cathode at a rate which initially obeyed Faraday's law of electrolysis. He also felt that hydrogen was absorbed by the anode.

5. Skinner obtained at least 2000 times as much hydrogen from a silver cathode than it could have contained. He felt that he could have obtained considerably more hydrogen, if he continued the experiment.

6. I propose that vacuum contains an arrangement of protons and electrons which can be converted into hydrogen by a process akin to electrolysis. I suggest a mechanism which, I feel explains Skinner's results. This mechanism predicts the shape of a curve given by Winchester for the production of hydrogen.

I have lost the rest of this letter and am not sure which article accompanied it. Pauling's reply follows:

Linus Pauling Institute of Science and Medicine
440 Page Mill Road, Palo Alto, California 94306
Telephone: (415 327-4060)
19 March 1985

Mr. Paul E. Rowe
34 Summit Avenue
Sharon, MA 02067

Dear Mr. Rowe:

I have read all of the material you sent to me with your letter of 22 February 1985. I return this material to you herewith.

First I must tell you that there have been extensive studies made of the absorption of gases on metal surfaces and other surfaces and on the solubility of hydrogen in various metals during recent years. I think that it is almost certain that the observations made by the early investigators are to be explained in terms of adsorption and absorption of hydrogen. For example, we know that sodium and potassium combine readily with hydrogen to form hydrides, which explains the absorption of hydrogen by these metals.

I have not found any of your ideas to be reasonable. I think that it is quite unlikely that these, which you have proposed only tentatively, have significance.

Sincerely,
Linus Pauling

The following is a copy my second letter and his second response. Don't bother to try to understand the tables in my letter. I hope they will make more sense after you read Part 2 of this article. I added it in 2004

The original letter included a drawing of the discharge tube. The apparatus employed in described in this copy.

34 Summit Avenue
Sharon, MA 02067
April 4, 1985

Professor Linus Pauling
Linus Pauling Institute of Science and Medicine
440 Page Hill Road
Palo Alto, CA 94306

Dear Professor Pauling:

I greatly appreciate your taking the time to read and comment on my paper. Your comments are, of course, quite valid. I am not angry and will continue vitamin C. I have been aware of the extensive studies of the absorption and adsorption of gases by metals and constantly refer to the 1962 edition of "Scientific foundations of Vacuum Technique", by Dushman and Lafferty. I believe that this book is considered the best summary of such information. I have also read some original articles on the subject, especially those pertaining

63

to the solution of hydrogen in silver at various temperatures. On page 427, Dushman makes the point that "activated adsorption has not been observed for the rare gases, for nitrogen on copper, or for hydrogen on gold and silver". The quantity of hydrogen Skinner reported obtaining from a tiny silver cathode could not have been in that cathode prior to his experiments. My experiments have employed copper, mercury and aluminum electrodes because they are available. I had to cease using copper cathodes since, in the gas produced by the discharge, a dark green transparent coat forms on the inside of the Pyrex discharge tube in the vicinity of the cathode tip. On being heated the coating evolves a gas and becomes a clear golden yellow and then a highly conductive copper colored foil. If the discharge is continued after the dark green coat has formed, a copper foil is formed directly around about a four-inch length of the tube nearest to the cathode tip. A clear brown coating forms on the glass around the next few inches of the cathode. The brown coating dissolves quickly in concentrated hydrochloric acid with the evolution of gas. On evaporation, a white solid is obtained. This is almost certainly cuprous chloride. I did not find reference to copper hydrides in recent articles on their structures, etc. The information most useful to me is in an 1899 textbook. A copy of part of the page is enclosed. I have little doubt that the yellow coating I obtained is the material referred to as cuprous hydride and the brown coat is that referred to as cupric hydride.

Data on page 528 of Dushman indicates that the solubility of hydrogen in copper is much too low to explain my observations. On page 427, however he gives a curve showing that considerable hydrogen is chemisorbed on copper in hydrogen atmospheres, I doubt that the quantities of hydrogen required to explain my results could have been retained by copper through extended exposure to atmospheric pressure and some vacuum exposure, but I have more work to do before I can make a definite statement. Enclosed please find a small piece of the copper foil and a small piece of glass with the brown coating.

I should point out that copper films are not expected to form under such conditions. According to Dushman (page 696, Table 10.1), at a pressure of 1 torr. Copper evaporation requires a temperature of 1622 °C. copper melts at 1084° C. The great portion of the copper film produced in my tube formed under a pressure of above 3 torr. of the gas produced by the discharge and a temperature in the range of 100° C.

The following experiment was performed using aluminum electrodes. On page 533, Dushman states that hydrogen is practically insoluble in solid aluminum, but is soluble to a small extent in the molten metal. He gives data on the solubility in the molten metal on the next page. On page 631, he gives a table which indicates that aluminum picks up no hydrogen when used as a getter. The apparatus I employed is pictured on the enclosed page. The discharge tube had been employed for several experiments over a period of two weeks and had not been opened to the air during that time. During this series of experiments wiring had been reversed several times, so both electrodes had been employed as cathode for considerable periods. It is well known that the properties of cathodes employed in low-pressure discharge change dramatically with use. The active ends of the electrodes had become quite rough and seemed to be covered with a dark gray non-metallic coating. I am enclosing the end of the anode which was employed in the following experiment. You may note that the portion of the electrode tip which touched or was close to the small diameter Pyrex tubing was practically unaffected by the discharge. The effects you see occurred when it was used as a cathode. The coating does not conduct electricity. All of the experiments performed in this tube were carried out in a helium atmosphere. Considerable gas was produced in each experiment. This was removed between experiments. The ends of the discharge tube were sealed with neoprene stoppers. The discharges were carried out under low amperage conditions and the various glows in the tube never came within 20 centimeters of the stoppers. The cathode was electrically grounded . It was felt many times throughout the experiments. At no time did the portion outside the tube become perceptibly warm. The glass tube was also felt throughout the experiments. Only the portion within about four centimeters of the cathode tip became noticeably warm.

Between experiments the leakage and or outgassing caused a pressure increase of about 2.5 microns per minute. This was probably due to leakage since the rate of pressure increase was practically independent of pressure between 10 and 4000 microns. Prior to the experiment discussed below the tube had not been used for 11 hours. The tube was evacuated to 27 microns and helium was added to a pressure of 2030 microns. As shown in the diagram, the less confined of the two electrodes was the cathode.

The experiment is preliminary in nature. Nothing is accurate. In order to show the effect of time between discharges on the voltage at which discharge starts, I had to increase the voltage, decrease the voltage and record the reading quite rapidly since the effect within the first minute after discharge seems very great. The time recorded is that at which I started to increase the voltage. The time to achieve discharge varied. Times between discharge listed in the table are therefore rough estimates. This was the first of several similar experiments. All of the experiments lead me to the following conclusions:

There is much that I don't understand about discharge tubes.

The shorter the time since the discharge stopped, the lower the voltage at which discharge will start. If the time between discharges is very short, discharge will start at very close to the voltage at which it stopped. The effect is such that even one minute is very important. The discharge can be started at a considerably lower voltage by tapping the tube in the vicinity of the cathode tip. Heating the tube in the vicinity of the cathode has a similar but not as dramatic effect. A coat builds up on the cathode tip during discharge. This appears to increase the voltage required to initiate the discharge.

Of course, these conclusions may not be valid for tubes with different electrodes or operated under different conditions. The second, third and fourth conclusions may indicate that something with a short half-life is formed during discharge and reduces the voltage required to produce a given current. Under each condition an equilibrium concentration of this something would eventually be achieved. When discharge is stopped, the concentration of this something would decrease with time.

This may explain the need for the series resistor which is employed to control the discharge. As the discharge starts, the discharge tube becomes more and more conductive. As this occurs the voltage drop across the resistor increases causing the voltage across the discharge to decrease. I have blown many fuses by using an insufficient series resistor.

Triatomic hydrogen may be the something referred to above. It was first discovered by J.J. Thomson in discharge tubes. Its existence was confirmed by Aston. Wendt and Landauer studied it extensively. They report that triatomic hydrogen, H_3, is unstable and reverts to H_2 in about a minute. A copy of part of the summary of their paper is enclosed. The fifth and sixth conclusions may corroborate my assumption that the reaction is initiated by the impact of a high-energy atom or molecule on the cathode surface. I can't help but think how I used to rap my radio which had a weak power tube to make it work.

I believe that the above results are consistent with the conclusions in my paper. However, Skinner's observation that hydrogen is initially produced at a rate consistent with Faraday's law of electrolysis is still the best evidence of an ether made up of protons and electrons.

I didn't intend to make this letter so long. I apologize. If you are at all interested, I would like permission to send you the results of my experiments every other month.

Very truly yours,
Paul E. Rowe

Enclosures:
Copper foil
Glass with brown coating
Cathode tip
Data from experiment
Photograph of apparatus employed in described experiments
Table A
(Please Skip this table and refer to it as you are directed in Part 2)

DATA FROM DISCHARGE EXPERIMENT USING
ALUMINUM ELECTRODES

Code	Time Minutes	Interval Minutes	Volts at Start	Pressure microns	Volts	Milliamps
A	0	--------	1480	2030		
A	1	¾	1240			
A	2	¾	1190			
A	3	¾	1130			
A	4	¾	1110			
A	5	¾	1110			
A	6	¾	1110			
	7			2100		
A	9	2 ¾	1150			
A	20	10 ¾	1350			
A	21	¾	1120			
A	22	¾	1130			
A	23	¾	1150			
A	24	¾	1170			
A	25	¾	1130			
26	26			2200		
B	27	1¾	1240		630	5.15
A	28 ½	½	1130			
	29			2290		
A	31	1	¾	1260		

32 Set at 1100v – no discharge. Started when bench on which the apparatus was assembled was tapped.

33 Set at 1050v - no discharge in one minute.

34 Heated cathode area of tube with a propane torch. Discharge started in approximately one second.

Code	Time Minutes	Interval Minutes	Volts at Start	Pressure microns	Volts	Milliamps
A	35	½	1040			

36 Heated cathode area for one minute.

Code	Time Minutes	Interval Minutes	Volts at Start	Pressure microns	Volts	Milliamps
A	37	1 ½	1250			
A	38	¾	1340			

Code	Time Minutes	Interval Minutes	Volts at Start	Pressure microns	Volts	Milliamps
A	39	¾	1260			
A	40	¾	1180			
	55		2400			
	56 Heated cathode area for one minute					
A	57	16 ¾	1200			
A	58	¾	1340			
A	59	¾	1320			
A	60	¾	1240			
A	61	¾	1160			
A	62	¾	1160			
A	63	¾	1140			
A	64	¾	1160			
A	65	¾	1170			
A	65 ½	¼	1080			
A	66	¼	1080			
A	66 ½	¼	1060			
A	67	¼	1070			
A	67 ¼	⅛	1010			
A	67½	⅛	1160			
A	68	¼	1130			
A	69	¾	970			
	70 Repeatedly increased and decreased voltage as rapidly as I could. Discharge restarted as low as 620 volts.					
A	72	?	1140			
A	73	¾	1180			
A	74	¾	1230			
A	75	¾	1260			
A	76	¾	1200			
	77		2500			
	78	1¾	1200	Let run. Set at 5 MA- Held at 5 MA		

Code	Time Minutes	Interval Minutes	Volts at Start	Pressure microns	Volts	Milliamps
	79			2560	634	5
	80			2610	626	5
	81			2660	622	5
	83			2680	612	5
	85			2700	----	----
	87			2700	603	5

88 Turned off and on quickly. Start at 607 volts. Turned variac down slowly. Off at 590 volts.

Code	Time Minutes	Interval Minutes	Volts at Start	Pressure microns	Volts	Milliamps
A	89		1250			
A	90	¾	1280			
A	91	¾	1360			
A	92	¾	1280			
A	93	¾	1260			
A	94	¾	1560			
A	95	¾	1230			
A	96	¾	1260			

97 Set at 1000 volts. Tapped lightly near cathode tip. Started on first tap.

98 Set at 1000 volts. Tapped lightly near anode – no discharge. Tapped harder near cathode. Started on first tap.

100 Set at 1000 volts. Tapped lightly near cathode – no discharge. Tapped harder– didn't start.

Code	Time Minutes	Interval Minutes	Volts at Start	Pressure microns	Volts	Milliamps
A	102	1¼	1380			
	103			2740		
	198			3000		
A	199	97		1590		
A	200	¾	1290			
A	201	¾	1240			
A	202	¾	1300			
A	203	¾	1250			

Code	Time Minutes	Interval Minutes	Volts at Start	Pressure microns	Volts	Milliamps

204 Continuously turned on and off every ¼ minute. (1120v, 1310v, 1350v, 1280v, 1380v, 1160v, 1340v, 1230v, 1220v, 1360v, 1200v, 1340v, 1320v)

Code	Time Minutes	Interval Minutes	Volts at Start	Pressure microns	Volts	Milliamps
207				3000		
431				3480		

Code A indicates that the voltage was increased from 0 to discharge over a period of about ten seconds.

Code B indicates that the discharge was started similarly but stopped after about one minute.

(The diagram in the letter is described below:)
A zero to 2000 volt DC power supply is connected to a discharge tube as follows:

The negative terminal is connected through a 100 kilo-ohm resistor and a milliamp meter to the grounded aluminum cathode. The positive terminal is connected directly to the aluminum anode. A kilovolt meter is connected between the cathode and the anode.

The body of the discharge tube is a Pyrex tube that has an inside diameter of 2.3 cm.

The cathode is a 3 mm. diameter aluminum rod. It passes through a neoprene stopper and is in the axial center of the tube. 37.4 cm. of the rod is in the discharge tube.

The anode is a 1.5 mm. Aluminum rod. It is centered in a 0.6 cm. inside diameter Pyrex tube that passes through a neoprene stopper. The anode passes through a small neoprene stopper. The inside end the anode tube is rounded and has a central 0.3 cm. hole, aligned with the cathode. The anode wire tip is 0.9 cm. inside the end of its Pyrex tube. It is 73.3 cm. from the inside of its larger stopper. The tips of the anode and the cathode are 10.0 cm. apart. The volume of the discharge tube is approximately 520 ml.

The following was copied from, "Inorganic Chemistry" by Ira Remsen 5th. ed. 1899

"Cuprous Hydride, CuH - This compound is made by treating a solution of barium hypophosphite with a solution of copper sulphate. It is thrown down as a yellow precipitate which gradually becomes darker. At 60° it decomposes into copper and hydrogen. With hydrochloric acid it yields cuprous chloride and hydrogen.

Cupric Hydride, CuH_2- is formed by treating a solution of copper sulfate with hypophosphorous acid. When freshly prepared it is a reddish-brown sponge-like mass, which however changes to a chocolate-colored powder on being freed from acid and boiled for some time. It is not readily changed when heated in air. It dissolves in hydrochloric acid with the evolution of hydrogen."

The following quote is from Gerald Wendt and Robert S. Landaur, JACS. 42, pp. 930-46 (1920)

"Summary

1. A reactive modification of hydrogen has been produced by several methods, all dependent on gaseous ionization- by the alpha rays from radium emanation, by the electrical discharge under reduced pressure, and by the high potential corona at atmospheric pressure. Attempts to produce activation by Schumann light rays failed.
2. This active hydrogen reduces sulfur, arsenic, phosphorous, mercury, nitrogen, and both acid and neutral permanganate. It is condensed or destroyed by liquid air temperatures. It is unstable and reverts to the ordinary form in about a minute. It passes readily through glass wool. It is not less stable at atmospheric pressure than at low pressures.
3. The activity is not due to gaseous ions, and the properties of the active gas are quite different from those of Langmuir's atomic hydrogen. The formation of a polyatomic molecule is indicated by the contraction of the hydrogen when ionized. Positive ray analysis at very low pressures shows a large proportion of triatomic hydrogen molecules which are likely the ones responsible for the chemical activity.

All the properties of the gas point to its being an ozone
form, perhaps properly called "hyzone."

(The following is Professor Pauling response to my second letter:)

LINUS PAULING INSTITUTE of SCIENCE and MEDICINE
440 Page Mill Road, Palo Alto. California 94306
Telephone, (415) 327-4064
19 April-1985

Mr. Paul E. Rowe
Summit Avenue
Sharon, MA 02067

Dear Mr. Rowe:
I have read your letter and examined the specimens you sent, and
have thought about your observations.

I am returning the specimens herewith, because I cannot think
of anything that I could do with them. Also, I must say that my
background of experience is such that I find it difficult to make any
suggestions to you. You may well have observed some interesting
phenomena, but I do not know enough to be able to discuss your
results in a meaningful way.

Sincerely,
Linus Pauling

PART 2

Oh well, he didn't call me doctor. I neglected to tell him that I
had a PhD. He was nice to respond and not call me a nut. I appreciate
his making the effort.

The following is my attempt to explain the results of the above
experiments assuming that vacuum contains a concentrated matrix
of protons and electrons:

As the discharge tube is evacuated, gases adsorbed on the electrode surfaces are removed.

As the voltage across the electrodes is increased, aether electrons are drawn closer to the anode surface and aether protons are drawn closer to the cathode surface. At some voltage, the most energetic collisions of gas atoms or molecules on the cathode supply sufficient energy to permit a cathode electron to combine with an adjacent aether proton to form a monatomic hydrogen atom, which is, then, released into the gas. This leaves an extra electron in the vicinity of the cathode. This electron is repelled from the cathode area and attracted toward the anode. This may initiate a falling domino- like effect that results in a neighboring aether electron being taken up by the anode. At this point, the rate of hydrogen production follows Faraday's law of electrolysis. (Note that this explains the ejection of electrons from the cathode at much lower voltages than occurs in much better vacuums.) Monatomic hydrogen is extremely reactive. It combines with the cathode surface to form a hydride, with another hydrogen atom to form diatomic hydrogen, with diatomic hydrogen to form triatomic hydrogen and possibly with the gas atoms or molecules originally in the tube. Triatomic hydrogen atoms diffuse through the tube. When they arrive at the anode, they deposit electrons and become triatomic hydrogen ions. These ions accelerate toward the cathode and pick up electrons to form triatomic hydrogen atoms.

The data in Part I of this paper suggests the triatomic hydrogen electron transfer is much more efficient than any other conduction mechanisms in the discharge tube. As the triatomic hydrogen concentration increases, the voltage across the discharge tube must be reduced or a fuse will blow (or worse).

At some voltage, much lower than that at discharge initiation, all glows disappear and the rate of triatomic hydrogen formation is the same as the rate of its dissociation. At that voltage, just enough hydrogen is produced to keep the triatomic hydrogen concentration constant. Very little of the current is due to electron transfer through the tube.

If the voltage is reduced the amperage drops and a new equilibrium concentration of triatomic hydrogen is achieved. At some voltage the discharge stops and no more triatomic hydrogen is produced and the triatomic hydrogen concentration quickly decreases.
The voltage at which the discharge will restart depends on how long it was stopped.

The following are conclusions from Table A:

Between 207 minutes and 431 minutes the pressure increases by 480 microns. This indicates a leakage-outgassing rate of 480/224 = 2.14 microns per minute. Some of this may be outgassing from the neoprene stoppers.

1. Between 79 minutes and 87 minutes, while the current was held at 5 milliamps, the voltage dropped from 634v to 603v and the pressure increased by 140microns. The rate of pressure increase between 79 and 81 minutes was (2660-2560)/2 = 50 microns per minute. This is almost 20 times the outgassing rate. At 87 minutes, the pressure leveled off.

I suspect this indicates the triatomic hydrogen concentration had approached the level required to stabilize at 5 milliamps under these conditions. At 32 minutes and at 97 minutes, the discharge did not start until the tube was tapped or shaken. This may have increased the velocity of some of the gas particles striking the cathode surface that started the discharge.

2. At 34 minutes the discharge started when he tube was heated in the vicinity of the cathode. This probably occurred for the same reason.

Paul E. Rowe 02649rowepaul@comcast.net
March 17, 2007

The following Is the portion of my play remaining, after I removed. "My Conversations with Einstein".

THE FALL AND RISE OF THE
HOUSE OF CARDS

Act I, Scene I
Explosives Laboratory, Hanover, MA, 1959

The curtain parts to reveal an almost empty stage. There is a lab bench at the center of the stage. On the bench is a vacuum pump, which is attached to a cylindrical steel chamber through a valve on the chamber. A mercury manometer is attached to another valve on the chamber.

At the left of the table is a steel plate about 7 feet high and 4 feet wide. Lined up against the back wall of the stage are many 2 5/8 by 3 1/2 foot playing cards.

At the top back center of the stage is a screen on which pertinent information can be projected. The words, "Explosives Laboratory, Hanover, Mass., 1959" are on the screen as the curtain opens. Pertinent information and/or diagrams will appear on the screen throughout the play, as required.

Enter Paul and his boss (Charlie). (As they approach the bench, they put on safety goggles.)

Paul: Last week I performed several experiments in this chamber using 20 mm. rounds loaded with various explosives. Prior to each explosion, I calculated the gas pressure I expected, based on the weight, chemical composition of the explosives and the gases expected. In each case, the pressure, when the chamber cooled, was quite close to the calculated value.

This week I have been testing the same explosives mixed with aluminum flake. Each experiment produced much more gas than is theoretically possible. At first, I employed just enough aluminum to react with the atoms in the explosive. When I reduced the explosive content and increased the aluminum content, I got much more gas. At first, I figured that gas adsorbed in the inside walls of the chamber, prior to the test, was released during the explosion. If this were the case, you'd expect to obtain less and less gas as more experiments were performed in the chamber. That didn't happen.

Charlie: Obviously there is something wrong with your analysis, Paul. Perhaps the extra gas is air that was adsorbed in the chamber walls between experiments.

Paul: I thought that might be the case, so I repeated one of the experiments but this time I heated the chamber to about 100 degrees C., while it was being evacuated and then evacuated the chamber for another hour as it cooled. This didn't reduce the amount of gas produced in the experiment.

Charlie: Well there is a simple explanation that has eluded you. Now, let me see if I understand the experiment you are about to show me. You made up three pellets of explosive-aluminum flake mixture and placed them in a 20 mm shell. You made one pellet of pure explosive, placed it on top of the other pellets and applied pressure to the top pellet to compact the assembly by the standard technique. You fixed a blasting cap to the top pellet and attached the blasting cap wires to the inside portion of the insulated electrodes that led through the chamber. The blasting cap wires are of such a length that the round is suspended vertically in the center of the chamber.

Paul: That's right. The pellets contain much more aluminum flake than should be required to react with all the chemicals in the shell and blasting cap. While the system was evacuating, the chamber was heated for an hour. Evacuation has been continuing as the chamber approached room temperature. You can see that the manometer indicates a perfect vacuum, which actually indicates a vacuum of 1 mm. of mercury or less. The valve to the vacuum pump has been closed for ten minutes and the reading hasn't changed; so there is very little leakage or out gassing. I am closing the valve to the manometer, to prevent mercury from being shot throughout the lab. (Charlie walks behind the steel barrier as Paul closes the valve). Now I'll attach this "ten shot" (picture of a ten-shot appears on the screen) to the outside portion of the electrodes. (Paul carries the "ten shot" with him behind the barrier.)

Paul (loudly.) Preparing to fire. (He waits 5 seconds.) Firing: Three—Two—One -- FIRE! (He twists the ten shot). A muffled boom is heard through the theater. (They walk to the bench and each of them carefully touches the chamber with a finger.)

Paul: See, Charlie. The explosion has increased the temperature considerably.

Charlie: Of course it has. I hope you don't expect me to hang around here while this thing cools down.

Paul: Certainly not! I'll get you in about a half an hour, when the chamber is cool enough that I can safely open the valve to the manometer.

Charlie: OK. (Charlie walks off left)

Paul: (Turning to the audience.) Charlie's the boss. I have to be nice to him. I'm sure you can entertain yourselves for the next half hour. (Walks over to the table and starts writing in his notebook. After five seconds. He scratches his head as if thinking. Looks at the audience.)

Paul: I was thinking. Time is supposed to be relative. I guess it is. It sneaks up on you when you least expect it. It's not so bad though. It never asks for money! (Looks at his watch.) Oh! Look! How time flies. I'll go get Charlie. (Walks off stage left.) (Returns with Charlie.)

Charlie: It can't be at room temperature already.

Paul: No it's slightly warm. It's safe to open the valve to the manometer, now. The pressure will be just slightly higher than it will be at equilibrium. I'll take a preliminary reading and the final reading in another hour. (Paul opens the valve and reads the manometer.) See!, we calculated that the pressure would be 193 mm. of mercury, but its actually 302 mm. I can't imagine what the extra gas can be, unless there Is a gas containing aluminum that no one has ever heard of.

Charlie: Well I admit to being stumped but I'm sure that you've made an error somewhere.

Paul: Most likely. If it's OK with you, I'd like to send a sample of the gas out for analysis.

Charlie: This project won't be renewed and there's very little money left. I can't let you have a mass spectrograph taken, but I'll ok a simple gas analysis.

Paul: Thanks. I'll send out three samples.

Charlie: Fine. (Charlie walks off left.)

Paul: (turns to audience) I suppose you don't want to wait two weeks; so I'll relativity up time, again. Oh. Look! The results just came in. (Picks up an envelope from the bench, opens it, removes a sheet of paper and reads it.) The relative quantities of oxygen, water, carbon monoxide, ammonia and carbon dioxide are reasonable. There are tiny traces of some organic compounds but the nitrogen content is way high. What's this asterisk? Let's see. (Reading) "We did not test for nitrogen. Any gas which is not among those listed was assumed to be nitrogen". (To the audience.) This is crazy! I can't account for much of the gas that they have included under nitrogen. This analysis doesn't explain the extra gas. I'm going away for a few years. While I'm gone, you try to figure it out. (Paul walks off stage left scratching his head.)

Curtain closes.

Act I, Scene II
Basement of Victorian House, Sharon, MA 1980

The curtain opens to reveal an almost bare stage. The large playing cards are still in the background. There are just enough props to suggest that the set is in a basement. Paul is sitting at a desk (center stage) entering data in his notebook. A bench (stage left) is set up for experimentation. There is a cot (stage right). Paul notices the audience. He faces the audience.

Paul: Oh! Hi. You missed quite a lot during the last 21 years. I'll try to bring you up to date on happenings that affect this play. Kate and I had three children. It's ok. She's my wife. She's great (Catherine the Great.) but she's not at all scientific. You probably won't meet her.

Well, anyway, having children and working with explosives didn't seem like a good idea. One clue was large extra payments for life insurance. So I changed jobs. The new job wasn't as exciting but it was lots safer and, to be honest, more interesting. I was involved in products mostly for the electronics industry. We developed plastic and ceramic materials that had special electrical properties. Some of the materials were sold to other manufacturers. We used many of the materials in producing products like radar absorbers or lenses and reflectors for radar. This forced me to learn something about electricity, magnetism, light and other aspects of physics.

I didn't tell you that my training was in chemistry. It was a great help in developing materials but I had to apply principals of physics to convert the materials to useful products.

I was still intrigued by the extra gas I had obtained in Scene I and wanted to find out what was going on. Since I had some lucky stock investments, I was able to take this year off. I spent part of the year in the MIT Science Library trying to find whether others had reported similar results. I have already learned that after World War II, articles in physics journals became more and more specialized and less and less readable. Luckily, most of the information I wanted was reported prior to World War II. Don't get me wrong. These articles require considerable study but, with effort, I could study them; sleep on them; read them again and then convince myself that I understood them. Please excuse me. I am about to enter the sleeping mode.

(Paul walks over to the cot. Lies down. Puts his head on a pillow and pulls up the covers.) (After 5 seconds, he lifts his head and turns to the audience.)

Paul: Oh! By the way, Catherine the Great says that I snore. I do not snore. If I fall asleep, I'll ask you to corroborate that when I awake. (Puts his head on the pillow. Falls asleep. Snores loudly.)

(During dream sequences, Paul may get out of the cot and move around the stage, as required.)

(The lighting on the stage becomes somewhat blue indicating a dream sequence. An early 20 th. century Englishman enters stage left, shakes Paul roughly and says) "Wake up! Wake Up." (Paul stops snoring and wakes up.)

Paul: What's the matter?

Englishman: You are snoring so loudly that you're disturbing the saints in heaven. St. Peter sent me down to shut you up.

Paul: Heaven? Who in hell are you?

Englishman: My name is J. Norman Collie.

Paul: If you were a Fellow of the Royal Society, I recently read a speech you gave to the society in 1913.

Collie: (Brightening up) Really! What did you think of it?

Paul: Well, to be honest. It was very interesting, but strange.

Collie: What did you find strange.

Paul: As I recall, you passed very high voltage electrical discharges through very low pressure gases and reported the appearance of gas atoms and molecules. In some experiments, helium appeared. In others helium and neon appeared. In some cases, hydrogen appeared, and in others disappeared. It was weird.

Collie: Let me point out that Patterson and also Masson reported similar results and we had all worked independently. After hearing about each other's results we published a paper together summarizing our results.

Paul: Yes I've read all those papers. I've also read two papers by scientists who were unable to repeat those results.

Collie: That is true. Actually, we reported that we couldn't always repeat our results. I was convinced that the results were correct but that some unknown variable needed to be controlled.

Paul: I came across those papers while trying to determine the source of unexpected gases I had obtained. For a while, I thought that the medium (aether), in which many scientists, of your day still believed carried light, might be helium. This would mean the whole knowable universe is permeated with helium. It pleased me to think that the most important thing in the universe was capital H small e, which in this form is the pronoun we use for God.

Collie: That's great! I'll tell Him! It will give Him a big kick. I have to leave now. Sleep quietly, for heaven's sake. (Collie leaves as he entered. Paul nods off and starts snoring, loudly.)

(A man walks on stage right. He is dressed like a 1920s professor of science. He walks over to and shakes Paul).

Man: Will you please wake up? (Paul wakes with a start and stops snoring)

Paul: What on earth are you doing?

Man: You mean, what am I doing on earth. Professor Collie sent me with instructions to keep you talking so you will stay awake. My name is Gerald Wendt.

Paul: Were you a professor at the University of Chicago?

Man: That's right.

Paul: What a coincidence. I recently read an interesting paper of yours in the 1922 Journal of the American Chemical Society.

Wendt: Do you really think that this is a coincidence.

Paul: (In a Loud whisper.) Quiet, I don't want the audience to learn that this play is rigged. As I recall, you and Clarence Irion exploded tungsten wires in one atmosphere pressure of carbon dioxide using high voltages and currents, in an effort to produce helium. After each explosion, you dissolved the carbon dioxide in a potassium hydroxide solution and collected the gas that remained. You tested the gas for possible decomposition products of carbon dioxide but all the tests were negative.

Wendt: That is correct. We intended to perform further tests on a 20 cc. sample of the gas that we kept, but the sample was lost through accident. We performed 21 experiments. The results were quite erratic because we couldn't get consistent explosions, but we obtained some of the gas in each explosion. The volume under standard conditions varied from 0.35 to 3.62 cc. The average volume was 1.01 cc., while the weight of the tungsten filament was 0.713 milligrams. If all of the tungsten atoms had disintegrated into helium, 4.0 cc. of helium would have been produced.

Paul: You would have an even harder time now, than you had then, convincing scientists that you had converted tungsten into helium. Did you, by any chance, test the gas for hydrogen?

Wendt: Why would we test for hydrogen? There was no hydrogen in the system.

Paul: When I detonated explosives containing aluminum flake in vacuum, I obtained much more gas than was theoretically possible and I am starting to wonder if that gas was hydrogen. There are similarities in our experiments. In each case, fine particles of molten metal were flying through a gas that contained carbon dioxide. Did you perform any further experiments along this line?

Wendt: No. I intended to, but I had some difficulty. Previous to the article being published, I had given an oral presentation to a meeting of the American Chemical Society. Regretfully, an exaggerated account of the speech was given wide publicity through the Associated Press. In any case, my health failed and I was told to rest completely for at least a year.

Paul: Yes. You included that information in a footnote. I doubt that you could include such personal information in a scientific article, today.

Wendt: I have to leave now. Professor Collie has been arranging for other visitors to keep you awake. I've enjoyed our dialog. Now don't go back to sleep. (As Wendt leaves another man enters. They nod to each other.)

Wendt: This is George Winchester of Washington and Jefferson College. I'm sure you will find him interesting.

Paul: (to Wendt) Goodbye and thank you. I certainly found your papers interesting. (To Winchester). Good Evening. I studied your article in the 1914 Physical Reviews. This is too strange to be coincidental. I suspect that Professor Collie is trying to help me.

Winchester: Of course, you are right. Since our deaths we have learned the truth about these matters, but we are not permitted to tell the living. We can however discuss anything we had published, within certain limits.

Paul: That seems fair. I will describe your experiments and hope you will stop me if I make an error.

Winchester: Go ahead. I'll enjoy this.

Paul: Here goes. You prepared glass tubes adapted with pure aluminum electrodes. There was a very short gap between the electrodes. While evacuating the system, you heated the tube with a flame to almost the softening temperature of the glass, in an effort to remove adsorbed gasses. When the tube cooled, you applied about 100,000 volts across the electrodes. The initial pressure in the tube was about one millionth of an atmosphere. You measured the pressure of the gas in the tube at various times and plotted the results. The pressure increased as the experiment progressed and continued to increase as long as the discharge continued. You analyzed the gas with a spectroscope. Initially the gas contained hydrogen, helium and neon. After a time, however only hydrogen was produced. You pointed out that hydrogen was evolved as long as long as any metal remained in the tube.

Winchester: Correct. Those conclusions were based on the results of several experiments performed under various conditions.

Paul: You suggested that the gases had probably been occluded in the metal electrodes. The helium and neon were probably near the surface of the metal and were released quickly, but hydrogen probably permeated the metal and was released over a longer period.

Winchester: Right.

Paul: Weren't you surprised at the quantity of hydrogen you obtained?

Winchester: Does it say that in my article?

Paul: No.

Winchester: I'm not allowed to comment.

Paul: How about a hint?

Winchester: Collie told you where we come from. Do you think we got there by disobeying the rules?

Paul: I suppose not. I won't press you further. Can you wait a minute? Nature is calling. (Paul leaves the stage).

(Winchester looks toward the audience)

Winchester: I didn't hear anything. Did you? Oh. I see! Well it's a long time since I was a food processor. You know, it is great knowing the answers, but it's frustrating not being able to help Paul more. However, I find that frustration isn't nearly as unpleasant as it used to be. (Paul returns).

Paul: Sorry to leave you like that. I hope you didn't give away the plot to the audience while I was gone.

Winchester: Don't worry. I didn't. There wasn't time.

Paul: I'll give them a clue. I don't expect you to confirm it.

Winchester: I won't.

Paul: I suspect that vacuum is not a void and whatever is in vacuum can be converted into hydrogen under surprisingly mild conditions.

Winchester: No comment.

(An English gentleman enters and pats Winchester on the back.)

Gentleman: Hello George. I am here to relieve you.

Winchester: Thank you J.J. Meet Paul. He just relieved himself. (To Paul) This is Sir J.J. Thomson. He can help you if anybody can. (Winchester leaves).

Paul: Thank you George. It was a pleasure.

Paul: (To Thomson). I am honored to meet one of the greatest experimenters in history. Let me tell the audience something about you. (Paul turns to the audience). The 1989 edition of the Encyclopaedia Britannica describes his work in a long article. It credits him with helping to revolutionize the knowledge of atomic structure by discovering the electron in 1896. He received the Nobel Prize for Physics in 1906 and was knighted in 1908. The article lists many other achievements including that he was an outstanding teacher and that his importance in physics depended almost as much on the work he inspired in others as on that which he did himself. Here is a direct quote from the article: (pretends to be reading an imaginary book.)

"Following the great discoveries of the 19th century in electricity, magnetism, and thermodynamics, many physicists in the 1880s were saying that their science was coming to an end like an exhausted mine. By 1900, however, only elderly conservatives held this view, and by 1914 a new physics was in existence, which raised, indeed, more questions than it could answer. The new physics was wildly exciting to those who, lucky enough to be engaged in it, saw its boundless possibilities. Probably not more than a half dozen great physicists were associated with this change. Although not everyone would have listed the same names, the majority of those qualified to judge would have included Thomson."

Thomson: You embarrass me. Are you sure that isn't my eulogy.

Paul: Yes. I'm sure. I am also sure that your work will continue to affect science for a long time to come. Anyway, you must have heard your own eulogy.

Thomson: No comment.

Paul: We'd better get down to business before you have to leave. I would like to discuss some of your work that led your colleague, Aston, to develop of the mass spectrometer. Here is my simplified description of the process: (Diagram appears on the screen.) You transferred gas at low pressure into an evacuated discharge tube. You increased the voltage between the electrodes until some gas particles in the tube absorbed enough energy to dissociate into positive ions and electrons. Some of these particles recombined releasing energy in the form of light. That caused the tube to glow. This is the principle of the, so-called, neon sign. You employed very low pressure, so some of the particles produced did not recombine. Electrons accelerated toward the positive electrode and positive ions accelerated toward the negative electrode, where they picked up electrons and formed atoms or molecules. Your negative electrode had a tiny hole that led, through a straight tube, to a second evacuated chamber. A small portion of the positive ions went through the hole in the electrode and headed into the second chamber. These ions passed through electric and magnetic fields that altered their paths. The heavier the ion and the smaller its charge, the less its path was altered. The second chamber contained a carefully positioned photographic plate. Ions that struck the plate caused a chemical change in the emulsion. You developed and analyzed the plate and were able to determine the weight and a rough concentration of each ion by the position and density of its affect on the plate. You referred to the ion stream as a positive ray.

Thomson: That's a good description. As you said, it was more complicated but that's the gist of it.

Paul: Thank you. (Takes out an imaginary sheet of paper) I have copied this quote from a 1920 article you wrote for Nature. You had pointed out that you were unable to obtain a plate in which the hydrogen line was absent. Here is the quote:

"I would like to direct attention to the analogy between the effect just described and an everyday experience with discharge tubes — I mean the difficulty of getting these tubes free from hydrogen when the test is made by a sensitive method like that of positive rays. Though you may heat the glass tube to the melting point, may dry the gases by liquid air or cooled charcoal, and free the gases you let into the tube as carefully as you will from hydrogen, you will get hydrogen lines by the positive ray method, even when the bulb has been running several hours a day for nearly a year."

Thomson: I admit that not being able to eliminate hydrogen bothered me greatly.

Paul: Did you consider that you might have somehow been producing hydrogen in the discharge tube?

Thomson: I am not allowed to answer that. I can say that I considered many causes for the hydrogen and rejected all of them. (A man enters the stage).

Thomson: It must be time for me to leave. Here comes Clarence Skinner. It is his turn to keep you awake.

Paul: Welcome, Professor Skinner. I would like to talk with Professor Thomson for just a minute before introducing you to the audience.

Skinner: If it's all right with him it's all right with me.

Thomson: Please be quick.

Paul: You found that the mass of the electron is about 1860 times less than the mass of the hydrogen atom.

Thomson. That is true.

Paul: It is believed that the hydrogen atom is made up of a proton and electron. So, the mass of the proton is about 1860 times the mass of the electron.

Thomson: I don't think any reputable physicist would disagree with that. (Thomson starts to leave.)

Thomson: Goodbye and good luck.

Paul: It was a great privilege discussing these matters with you.

Skinner: There goes a wonderful gentleman.

Paul: I agree, but you are the person I most wanted to meet. (To the audience.) This is Professor Clarence A. Skinner of the University of Nebraska. He performed many very interesting electrical discharge experiments early in the twentieth century. I particularly want to discuss results he reported in the Physical Review of 1905.

Skinner: That article was referred to in many of the papers you discussed previously.

Paul: I know. That's how I found it. It includes many surprising experimental results. As usual, I will try to summarize what interested me most. (Picks up an imaginary paper.) This is a copy of your paper. I will read the first two paragraphs.

"While making an experimental study of the cathode fall (voltage drop at the negative electrode) of various metals in helium it was observed that no matter how carefully the gas was purified the hydrogen radiation, tested spectroscopically, persistently appeared in the cathode (negative electrode) glow. Simultaneously with this appearance there was also a continuous increase in the gas pressure with time of discharge. This change in gas pressure was remarkable because of its being much greater than that which had been observed under the same conditions with nitrogen, oxygen or hydrogen. Now the variation in the cathode fall with current density and with gas pressure in helium was found to be so like that obtained earlier with hydrogen that it appeared necessary to maintain the helium free of the latter in order to make sure that the hydrogen present was not the factor causing this similarity in the results. Futile endeavors to attain this condition led to the present investigation, which locates the source of hydrogen in the cathode, shows that the quantity of hydrogen evolved by a fresh cathode obeys Faraday's law for electrolysis, and that a fresh anode (positive electrode) absorbs hydrogen according to the same law."

Skinner: Yes. I wrote that, except for your three asides.

Paul: First let's discuss the effect you noted with the different gases. This quote is from the body of your paper:

"If now hydrogen is liberated from the cathode in all gases we should expect in nitrogen an increase in gas pressure arising from the formation of ammonia, which takes place when a discharge passes through a mixture of nitrogen and hydrogen. In this case the rate of increase in gas pressure, compared with that in helium, is small since six atoms of hydrogen would be required to change one molecule of nitrogen into two molecules of ammonia, while in chemically inactive helium one new molecule of gas is formed from two atoms of hydrogen. Likewise with oxygen water vapor would be formed and absorbed in the dryer, in which case, four atoms of hydrogen would cause a decrease of one molecule in the gas filling."

Paul: I quoted this part because it confirms that the gas you produced was, indeed, hydrogen.

Skinner: I agree, that is the reason it was included in the article. Why didn't you finish the paragraph?

Paul: Because I don't agree with your explanation for the lack of pressure increase in hydrogen. It is my play and I intend to keep control of it.

Skinner: Well, I would like to hear your ideas on this matter.

Paul: I would be glad to discuss this some other time. You must realize that most members of the audience do not have a scientific background. I'm probably losing their attention as it is.

Skinner: All right. It's your play. Continue.

Paul: The original quote stated that hydrogen was evolved at the cathode and that the initial rate of evolution obeyed Faraday's law of electrolysis. That suggests that for each electron that leaves the cathode, a hydrogen atom is formed. The same is true when hydrogen is produced by the electrolysis of water. In the case of water, the hydrogen comes from the water. We don't believe that there was any hydrogen in your tube. If there were free protons in the tube, hydrogen would be produced at the rate you found. Do you have a comment?

Skinner: Only that such an idea would not have been accepted in 1905. At that time, I believed that the hydrogen was present in the cathode and that the electrical conditions caused the hydrogen to be released into the discharge tube.

Paul: My suggestion is not popular today, either. I am surprised that you didn't stress your observation that tarnish appeared on the surface of all metal cathodes during your experiments. This is further evidence for production of a gas other than helium during discharge.

Skinner: I thought that was quite obvious.

Paul: Here is another quote from your paper:

"An endeavor was made to deplete the metals of their supply of hydrogen by passing from them as cathode a current for a sufficient length of time, but after a time this rate begins to drop off until the pressure appears to have reached a constant maximum value. Silver was depleted in this way giving off about two tenths of a cubic centimeter (measured at atmospheric pressure) of hydrogen. The current was then broken and the hydrogen absorbed by the Na,K (Sodium, Potassium) cathode. After standing in helium overnight then tested again the next morning it was found to have a new supply equal to the one given up the day before. Without allowing it any chance of regaining hydrogen from an external source it was thus

90

depleted six or eight times during the course of two weeks and found to give off at each time about the same amount of gas."

Did you believe that your silver cathode could have contained that much hydrogen?

Skinner: I'll admit to having been surprised. I was convinced that the hydrogen was produced at the cathode and that the helium in the tube was pure. It seemed to be the only reasonable conclusion.

Paul: Here is a quote discussing the same experiment:

"Altogether about two cubic centimeters of gas have been given off by this silver disk, which is 15 mm. in diameter and about one mm. thick. It shows no sign of having its supply of hydrogen reduced in the least." Did you believe that you could obtain an infinite quantity of hydrogen if you continued the experiment indefinitely.

Skinner: No, but I believed that the silver disk contained much more hydrogen than I had removed.

Paul: I can understand that. Let me discuss an article in the 1928 Proceedings of the Royal Society, London. It was written by E.W.R. Stearcie and F. M. G. Johnson of McGill University and titled, "The Solubility of Hydrogen in Silver". It is an exhaustive study of the subject. Based on their findings, I calculated that you produced thousands of times more hydrogen from that silver cathode than it could have contained.

Skinner: Well that is very interesting. I would like to comment, but I can't. Anyway, it's getting light out and I have to leave.

Paul: I'd like to make the following comment: You mentioned that, on standing overnight between these experiments, the tarnish fell from the silver cathode leaving a clean metallic surface. Did you know that silver hydride is unstable and soon decomposes into silver and hydrogen.

Skinner: I can't answer that.

Paul: I am sorry we don't have more time. Your paper is the most interesting that I ever studied. There is a lot more in it that I would like to discuss.

Skinner: It will have to wait, unless you are dying to discuss it.

Paul: I think I'll wait. Your results convinced me that there is something in vacuum that can be converted into hydrogen gas.

Skinner: No comment (Skinner leaves and Paul returns to the cot and snores so loudly that he wakes himself. (The light becomes normal again.)

Paul: (To audience, remembers where he is.) See, I told you that I don't snore.

(Curtain closes on Act I, Scene II)

Act I, Scene III
Basement of a new house, Mashpee, MA, 1996

The curtain opens to reveal a basement with two lab benches, a desk, a cot and several pieces of lab equipment. The large cards and the projection screen are still in the background. Paul is sitting at the desk. He scans the audience and appears concerned.

Paul: (To audience Welcome back. I'm confused! Am I the only one who has aged in the last 37 years. A few years ago, Kate and I moved this new house on Cape Cod. This is my new laborato …

Kate: (off stage) Paul. Mort and his friends are here.

Paul: I knew Kate would work her way into my play, somehow. Excuse me.

Paul: (To Kate) Great! Send them down. (Enter Mort, Curley and Larry). (Mort looks like and is a distinguished businessman. Curly and Larry are also successful but still, somehow, resemble two of the three stooges. They didn't, but it is time for some comic relief and I forget their real names.)

Mort: Paul, meet two of my college classmates, Curley and Larry Zart. We became reacquainted yesterday at our 50th MIT reunion.

Paul: Pleased to meet you. Where is Moe.

Curley: Hello Paul. Do you know our brother Moe?

Paul: No. I was just fooling. Everyone admires Moe Zart.

Larry: (Grabbing and pulling Curley's nose with one hand, hitting his own hand with his other hand. (This results in a usual 3 Stooges noise). That was a joke, Curley. (He shakes one of his hands in disgust.) Anybody got a handkerchief. (Paul takes a roll of paper towels from a bench, tears off a sheet and hands it to Larry. Larry wipes his hands and throws the paper into a wastebasket). (To Curley) You're such a slob.

Mort: Stop playing around. If you pay attention, Paul will show you a very interesting experiment.

Larry: I'm already impressed with this laboratory. How did you get all of this equipment?

Paul: I got much of it from sealed bid auctions of Raytheon's surplus equipment. The bids were quite low on most large complicated pieces of equipment. For example, I got this portable lab bench equipped with a great vacuum pump, two steel vacuum chambers and a large pressure chamber for $212.00. Of course, I had to disassemble many large items to get the gauges, stopcocks, etc., I needed. I got 16 boxes of 4 foot long, 1 inch diameter glass tubing for $22.

Curley: How about all this other stuff.

Paul: Much of it came from Raytheon or Polaroid surplus sales. Some was discarded where I worked. I even got some at yard and estate sales. As a last resort, I paid full price at chemical or electrical supply houses. I almost purchased an old fashioned mass spectrometer from Union Carbide. I bid $ 1000.00 based on a picture and description in their auction catalogue. This was much more than I paid for anything else. They informed me that I was high bidder, asked me how I wanted it shipped from West Virginia and informed me that I would need a crane to remove two extremely heavy electromagnets from the truck. There was no way I was going to get those magnets in this basement. I figured, my car would never be in the garage again. For better or for worse, the Carbide agent called and apologized. When they went to crate the equipment for shipping, they found that it had been savaged for parts and would never be operational. I had mixed feelings but the main feeling was relief.

Mort : You were lucky. Kate would have killed you.

Paul: I doubt it. I'm underinsured.

Mort: Let's get down to business. Tell Larry and Curley what you're going to show them.

Paul: Sure. I've set up an experiment that will produce considerable hydrogen gas by combusting a mixture of pure aluminum and copper oxide in vacuum.

Larry: Come on, Paul. That's ridiculous. You can't produce hydrogen from aluminum and copper oxide.

Paul: I agree.

Larry: But you just said …

Paul: That's right. I suspect that vacuum is not a void. It must contain something that can be converted into hydrogen, under the proper conditions. That is what I am trying to show you.

Curley: Good luck. You'll need it.

Paul: Thank you, Curley. Perhaps you will do me a favor. (Hands Curley a video camera) Please record the experiment as we perform it.

Curley: I'll do my best. Just remember, I'm not a pro.

Paul: I'm sure you'll do fine. (To the audience) It's not important. I intended to include detailed verbal descriptions and pictures of me performing this and another experiment, but it took too much time. I feared that you would miss the last bus or get in trouble with your baby sitters parents. The screen will show appropriate information for those of you with a scientific background.

Larry: You will have a hard time convincing me that you can produce hydrogen in this set up.

Paul: At least you are giving me the chance. I submitted papers describing this and other experiments to respected scientific journals, but they were rejected. The Journal of the American Chemical Society rejected a paper. They wrote that it contained references to old literature and nothing new. I suspect my papers were rejected because they don't conform to current theories. My Thermite type experiment is over here. (Group moves to the lab bench.) Here's how I set it up. (Appropriate pictures and equations appear on the screen.)

I prepared a glass test tube for the experiment. I added 15.3 grams of cupric oxide. Its purity is better than 98 %. This gives similar results to the CP (chemically pure) cupric oxide I used in earlier experiments, but cost much less. I added 8.0 grams of 99.5+ % pure aluminum powder from Alcoa and poured the contents of the tube onto a carefully cleaned 20 mesh screen which was resting on a piece of white notebook paper. I shook the screen until all of the powder passed through. Using a stainless steel spatula, I blended the powders. The screening and blending procedure was repeated three times to produce an evenly colored mixture. After bending the paper into a trough, I poured the mix back into the tube and weighed the assembly to determine the weight of the mixture by subtracting the original weight of the tube. As you can see by examining the paper that I used, the loss is negligible.

Larry: Don't you think that some paper might have gotten into the mixture?

Paul: Examine the paper. What do you think?

Larry: It doesn't look like it but …

Paul: You are right to suggest the possibility. Experiments where I mixed the ingredients on a polyethylene sheet or on glazed ceramic tiles produced similar results.

Larry: Good, but I'll keep looking for errors

Paul: Please do. That's the whole idea. I previously prepared a Cromel wire coil, attached copper wires to both ends and forced the coil into the mixture. I fixed the coil in place by bending the lead wires around two glass ears on the side of the glass tube and attached the open ends of the wire to two of the electrodes that pass through the vacuum chamber cover. I had previously attached a similar coil to the two other electrodes in the cover. The cover was then clamped onto the chamber. The glass tube is suspended in the middle of the chamber. The second Chromel coil is inside the chamber, near the cover and will be exposed to any gas that may be produced in the chamber. Chromel is a metal resistance wire that is often used in electrical heaters. It is, likely, used in the appliance you use to keep your remaining coffee warm, while you are having your first cup.

Larry: Are you sure that the gas you produce doesn't come from the Chromel?

Paul: Almost certain. Varying the length or diameter of the wire has a negligible effect on the amount of gas produced in these experiments. This indicates that the gas doesn't come from the Chromel wire.

The chamber was heated to 80 degrees C. for an hour, while being evacuated. This should have removed some of the gas that might have been absorbed in the system. I have found that whether the chamber is heated or not has little effect on the quantity of gas produced in the experiments. As the pressure decreased, the mercury level dropped in the sealed leg of the manometer, and rose in the evacuated leg. The difference in levels is the pressure in torr. Torr. is a unit of pressure formerly known as millimeter of mercury. It is about 1/760th of atmospheric pressure.

Mort: Is this a good time to tell us how you got involved in this?

Paul: Sure. I'll be brief, since I'm repeating what that the audience was exposed to in the beginning of Act I. After 37 years, a little reminding shouldn't hurt. I obtained more gas than is theoretically possible, when I detonated aluminum flake containing explosives in vacuum. An exhaustive literature search convinced me that the extra gas was hydrogen. The explosives I used were fairly complicated chemicals that contained some hydrogen atoms. (To audience.) You can pay attention again. Here's some new information. (To actors.) I found that a simpler combination, Thermite, a mixture of aluminum and rouge produced hydrogen gas when combusted in vacuum. (To audience.) As all you ladies know, rouge is finely powdered ferric oxide. (To actors.) I tested many aluminum-metal oxide mixtures. The easiest to ignite was fine aluminum powder and cupric oxide. A mixture of pure aluminum powder and CP (chemically pure) cupric oxide (Baker Analyzed) produced more hydrogen than any other mix I tested. Mixtures of equal total weight that contained 50 % more aluminum than is required to react with the cupric oxide, produced considerably more hydrogen, than a mixture calculated to give complete reaction between cupric oxide and aluminum. I have performed that experiment many times and found that the results are quite reproducible. Today I am performing a version of the experiment, for you.

I also repeated some of Skinner's (To audience.) We met him earlier. (To actors.) I performed electrical discharge experiments and convinced myself that he had, as he stated, produced hydrogen in his discharge tubes. The quantities of hydrogen I obtained were many times greater than could have been present prior to the experiments. I even produced hydrogen in clear, evacuated Pyrex tubes that were adjacent to operating high voltage spark coils. This is known as electrodeless discharge. There was no metal in the tubes, only very low pressure gas and Pyrex. Such tubes produced no hydrogen when under extremely low pressure or when filled with helium at low pressure. They produced hydrogen when a trace of oxygen or a gas whose molecules contain oxygen atoms was present in the helium. When very low pressure oxygen was present, the pressure increased during discharge. When the gas produced was exposed to drying agents the pressure fell indicating that oxygen had reacted with hydrogen produced by the discharge to form water.

Mort: Thanks Paul. Enough talk. Let's see the experiment.

Paul: I'm ready, if you are.

Curly: Go to it.

Paul: First, I would like each of you to touch the reaction chamber and tell me how it feels. (They do so).

Larry: It feels quite cool.

Curley: I agree.

Mort: Me, too. It feels just as you would expect. It is at room temperature but feels somewhat cooler because the metal draws heat from your hand.

Paul: The system has been evacuating with the valve to the gauge system open. There are two measuring devises in the gauge system. A McLeod gauge records pressures between 0.001 and 5 torr. The manometer records pressures between 1 and 200 torr. At this time, the McLeod gauge indicates a pressure of 0.260 torr. I will close the valve to the vacuum pump and allow the system to leak and/or outgas for 30 minutes. The audience has allowed me to speed time in earlier. Why not now? Well, it's time to take another reading. Let's see. The McLeod gauge reads 1.550 torr., indicating an increase in pressure of 0.042 torr. per minute. In the vacuum business this is quite substantial, but it is extremely small compared to the quantity of gas that will be produced. I am opening the valve to the vacuum pump and evacuating for 4.5 minutes How time flies! Let's take another reading. The pressure is 0.355 torr. In order to isolate the reaction chamber, I am closing the valve to the vacuum pump and the valve to the gauge system. Now, I am applying a current through the Chromel coil with this variable transformer. This causes the coil to become red hot. (A loud puff is heard). That noise indicates that the heated coil has caused a reaction. Now feel the steel chamber. (They do).

Larry: It is quite warm.

Curley: I'd call it hot.

Paul: The reaction of cupric oxide and aluminum to form copper and aluminum oxide produces lots of heat. The steel chamber is quite massive. It's impressive that such a small quantity of powder can increase the temperature of the chamber this much . Let's speed time again … Sixty minutes have passed (Feels the chamber.) and the chamber is cool. Keep an eye on the mercury manometer while I open the valve to the gauge system. (Opens the valve.) See how the mercury level has changed. This indicates that considerable gas has been produced. (He reads the manometer.) The pressure has increased from less than 1 torr. to 78 torr. (He calculates on a sheet of paper.) Let's see. The volume of the system is 3 liters: so the gas produced would have a volume of 19.8 cubic centimeters at atmospheric pressure and room temperature. A 25 cent piece has a volume of about 1 cubic centimeter. So the gas produced has the volume of a stack of 20 quarters.

Mort: How much gas would you have gotten if the hot coil had decomposed all the cupric oxide into copper and oxygen gas?

Paul: Good question. It would be of the same order of magnitude. If that had happened, however, there would have been no noise and the chamber would have remained cool. The next step should convince you that the gas in the chamber is hydrogen. I am removing the electrical leads from the electrodes to the used coil and attaching them to the coil near the top of the chamber. (He does so.) Now I am increasing the setting on the variable transformer to a position that will cause the coil to become red hot. Note that there is no noise. Now feel the chamber. (They do.)

Mort: I don't feel any change.

Larry: Agreed.

Curly: I guess this is confirmation that the heat we felt earlier wasn't from the Chromel coil.

Paul: And that the gas in the chamber does not react further under these conditions. Now I am carefully opening this valve slightly and quickly closing it, to let a small amount of air into the chamber. (He reads the manometer). The pressure in the chamber is now 227 torr. This indicates that chamber contains 78 torr. of the gas we produced and 149 torr. of air. Now I will close the valve to the gauge system (does so.) and turn the variable transformer to the same setting as previously. (He does so. A ping is heard.) That ping indicates that the gas we produced reacted with air. Feel the chamber again. (They do so.)

Larry: It is definitely warm but much cooler than after the earlier reaction.

Curley: I agree. I guess you think that the original gas was hydrogen and that it reacted with air to produce water.

Paul: Exactly. I will calculate what the final pressure should be, on that basis. (He uses a calculator and a piece of paper) According to this there was insufficient oxygen to react with all the hydrogen. The chamber should contain the following:

18.4 torr. of hydrogen, 119.0 torr of nitrogen and 59.6 torr. of water. (Looks in a handbook) According to this table any gaseous water over 23.8 torr. will liquefy at this temperature. The pressure should be 18.4 + 119.0 + 23.8 or 161.2 torr. Let's speed up time again while the system equilibrates. The chamber is cool now. I will open the valve to the gauge system and read the manometer. (He does so) The pressure is 163.0 torr., which agrees well within experimental error with my calculated value. The original gas must have been hydrogen. No other gas would give this result.

I should point out that I have performed this experiment in clear Pyrex flasks using less mixture. The conclusions were the same and the gases produced had no color and no odor. This is consistent with the formation of hydrogen.

Larry: What if you added a different amount of air to the gas you produced.

Paul: I'm glad you asked. My technique for adding air is quite crude and, therefore, not reproducible. I have performed this experiment with various excesses of oxygen and various excesses of the reaction gas. The results always led me to the same conclusion. I have no doubt that the gas is hydrogen.

Larry: As you know, our background is in electrical engineering. Our knowledge of chemistry is limited. Your experiments seem quite straight forward and extremely interesting, but I'm not qualified to comment further.

Paul: I understand. Since your background is electrical, perhaps you can explain the forces between these two magnets. Takes a pair of magnets from a drawer and hands them to Larry. Larry manipulates them with interest.

Larry: I believe it has to do with magnetic fields and lines of force.

Paul: Do you really believe that two magnets separated by vacuum can attract or repel each other, if vacuum is a void? Doesn't that imply that the void in the region of the magnets is different than the void far from the magnets?

Larry: I admit that my explanation doesn't make sense. I had trouble with it in school, but, at some point, one has to accept such concepts and get on with his education.

Curley: You would have to be an Einstein to understand that.

Paul: Or to convince everybody that you understood it. Let's go upstairs. Kate has a snack for you. If you agree with my ideas, I'll get you a drink. (They chuckle and leave the stage.)

Curtain falls

Act II Scene I
Same stage setting. (1999)

Curtain rises to show Paul sitting at the desk, writing in his notebook. He turns to the audience.

Paul: Welcome back. Let's have a show of hands. How many of you are completely confused by this play? Sorry about that! Now, is there anyone here who follows everything, understands completely, and believes every word? Well, are any of you actors impressed? Ok. Perhaps a short summary is in order:

1. I obtained much more gas than was theoretically possible by detonating explosives containing aluminum in vacuum.
2. A literature search revealed that many respected experimenters, some quite famous, reported obtaining surprising amounts of hydrogen gas in their experiments.
3. I produced hydrogen gas by reacting mixtures of cupric oxide and aluminum powder in vacuum.

4. Like Skinner, I produced hydrogen gas during electrical discharge in low, pressure helium. I also produced hydrogen in a fairly good vacuum.
5. I produced hydrogen gas by other techniques, including placing a glass tube containing a fairly good vacuum near an operating spark coil.

This has led me to believe that vacuum is not a void. It contains something that can be converted into hydrogen, under the proper conditions. My best guess, so far, is that the knowable universe is permeated with a matrix of protons and electrons. This may be aether, the light carrying medium, that scientist accepted as fact, in the late 19th and early 20th centuries. If this is the case, why isn't it obvious to everyone? How can I move my hand through such a matrix with such little effort? I performed another literature search in an effort to answer such questions. As usual, I will take a nap and let my unconscious mind work on it.

(Paul walks over to the cot and lays on it.) Now I can consider the problem in perfect silence. (Paul lays his head on the pillow, closes his eyes and snores. The bluish light covers the stage indicating dreaming.) (A 17th century Dutch gentleman enters walks over to Paul and shakes him. Paul opens his eyes looks at the gentleman and says)

Paul: I must be dreaming.

Gentleman: Yes, but you are no longer snoring.

Paul: You appear to be from the distant past.

Gentleman: That is correct. I lived in 17th century Holland. My name is Christiaan Huygens.

Paul: It is a pleasure to meet such a famous scientist. It seems to me your conclusions are much more reasonable than those accepted today. I'm surprised that you speak English.

Huygens: Everyone in heaven speaks English, since England conquered heaven under Elisabeth I.

Paul: Did Professor Collie send you?

Huygens: Yes. Both Collie and St. Peter.

Paul: Then, I expect that you won't stay very long. I'd like to summarize part of one of your books.

Huygens: Fine. Go ahead.

Paul: Your "Treatise on Light" was published in 1690. I read an English translation in "Great Books of the Western World". You proposed that light (like sound) is a wave phenomenon transported through a medium of material particles. You suggested that an evacuated bell jar does not carry sound because the medium for sound transmission is missing. Since light is transmitted, its medium must still be present.

Huygens: That is what I believed.

Paul: And now?

Huygens: You should know that I may not answer that.

Paul: It was worth a try. You discussed an experiment of Torricelli, a contemporary of Galileo. Both lived in Pisa. I guess they were Pisanos.

Huygens: Are you trying to entertain the Mafia?

Paul: No. Just a few Sopranos.

Huygens: Enough useless banter. Let's get serious.

Paul: OK. Here goes. Torrecelli filled a glass U-tube with mercury to a sealed end and evacuated it from the open end. You noted that the space that developed between the mercury and the sealed end transmitted light. On this basis, you concluded that the medium for light was still present and that the medium easily passes through spaces between the atoms of the glass and/or the mercury. You pointed out that this indicated that the particles that make up the medium must be very small, indeed. You considered light to be transferred by a mechanism similar to the transfer of energy from ball to ball in a series of suspended metal balls. At any time, all of the energy is on only one ball. Energy is transferred from ball to neighboring ball. The velocity of energy transfer depends on the physical properties of the material of the balls. Here is a quote from your book. (Reading)

"And it must be known that, although the particles of the ether are not ranged thus in straight lines, as in our row of spheres, but confusedly, so that one of them touches several others. This does not hinder them from transmitting their movement and spreading it always forward."

You assumed that each activated aether particle passed all of its energy to neighboring aether particles. That is, each activated aether particle is the start of a new wave. On this basis, you developed equations that predict observed diffraction patterns. For many years, this was considered strong evidence for a material aether.

Huygens: I believed that my calculations wouldn't have predicted the various diffraction patterns if such a medium wasn't present. The scientific community agreed for two centuries. You should have defined diffraction patterns for the audience. (Faces the audience) When light from a single point source falls on an opaque plate which has small openings, for example, two parallel slits, light that passes through the slits onto a dark plate parallel to the opaque plate and some distance from the slits, forms a series of light and dark lines on the dark plate. I was able to derive a formula that predicts the pattern of lines and how they would vary as the distance between the slits and/or the distance between the plates varies. That is a simple example of a diffraction pattern.

Paul: That was very good. I wish I'd said that.

Huygens: Don't worry. You will.

Paul: I'll consider it … Are you suggesting that I reduce your role in this play?

Huygens: Perhaps. Am I being paid by the word?

Paul: We'll discuss this off stage. Let's get on with the play. If you had added that each activated particle passed all of its energy to one and only one, neighboring aether particle, your aether concept might still be accepted. After your death, more information about light became available. If energy from an active aether particle was passed to more than one particle the frequency of the light would decrease rapidly as the light moved through the medium. This is certainly not the case. If each active particle passed all of its energy to only one adjacent particle the frequency would not change with distance traveled. Each activated aether particle (or group of particles) carrying a given frequency would have the same energy and would produce the same effects, regardless of the distance from the source of energy. That is, each photon transmitting a given frequency would have the same energy. There would simply be fewer photons per volume as the light traveled. This provides a very simple explanation for the photoelectric effect for which Albert Einstein received his Nobel Prize. It would not be possible to predict which neighboring particle would become energized by an active aether particle. This may bear similarity to the Heisenberg uncertainty principle.

Huygens: What is a Nobel Prize?

Paul: A man named Nobel left a fortune to be invested. Each year, outstanding individuals in various fields are given large cash prizes. You would have liked Alfred Nobel. He was a dynamite guy. Fesnel effects, discovered many years after your death, were found to conform to your equations. This was considered further confirmation of your material aether.

Huygens: As you know, I can't comment. Keep up the good work. Perhaps you will add to my reputation. (A gentleman from the late 19th century enters and taps Huygens on the shoulder.)

Gentleman: Hi Chris. I'm here to send you back. How is it going?

Huygens: Quite well, thank you. Paul this is James Clerk Maxwell. He'll enjoy discussing these matters. It's time for me to leave. (To Paul) Good luck and keep the noise down. (Huygens starts to leave.)

Maxwell: (To Huygens.) I'll try to keep him from snoring. (To Paul.) It's nice to meet you, Paul. I hear you have some unpopular ideas.

Paul: That's true. My ideas are not appreciated these days. I think they would have been considered reasonable in your day. Oh! Pardon me. Let me introduce you to the audience. (Turns to audience.) This is Professor James Clerk Maxwell. According to the 1989 edition of the Encyclopedia Britannica, (Reading.) "He is often ranked with Sir Isaac Newton for the fundamental nature of his contributions to Science". (To Maxwell) It is an honor to meet you sir. Is it true that you manufactured Jack Benny's car?

Maxwell: Is this going to be a serious conversation?

Paul: I hope so. I'm sorry, but every once in a while I feel the necessity of waking the audience. We don't want them to snore.

Maxwell: You're excused. What would you like to discuss?

Paul: In the late 19th century, you developed the equations we still employ to predict the behavior of electromagnetic radiation of all frequencies and in all materials. In developing the equations, you assumed the presence of a concentrated, material aether. The following three quotes are from your book on the subject: (Reading.) Here's the first quote:

"In several parts of this treatise an attempt has been made to explain electromagnetic phenomena by means of mechanical action transmitted from one body to another by means of a medium occupying the space between them." Here's the next quote:

"According to the theory of undulation, there is a material medium which fills the space between the two bodies and it is by the action of contiguous parts of this medium that the energy is passed on, from one portion to the next, til it reaches the illuminated body." And the third quote:

"Let us next determine the conditions of propagation of an electromagnetic disturbance through a uniform medium, which we shall suppose to be at rest, that is, to have no motion except that which may be involved in electromagnetic disturbances. Let c be the specific conductivity of the medium, k its specific capacity and u its magnetic permeability."

Paul: So, like Huygens, you believed in a medium of particles that carries light as a wave phenomenon and that the particles of this medium are contiguous, that is, touching. On this basis and using known values, including the speed of light, you calculated the permittivity (e_0) and permeability (u_0) of vacuum.

Maxwell: You've summarized the book very well. Of course, I used some fairly deep mathematics. If you included that, there would have been a chorus of snores from the audience.

Paul: I'm glad you are getting into the spirit of the play. Permeability has to do with magnetism and is now associated with unpaired electrons.

Maxwell: J.J. Thomson determined the mass of the electron soon after my book was published. He greatly advanced science with his discoveries.

Paul: Your wave equations are used today to successfully calculate the behavior of electromagnetic waves of all frequencies, some of which you didn't know existed. Did you have any doubt that light was transferred through a material medium?

Maxwell: At the time, it was generally accepted that such a medium existed. I considered my equations further proof of the concept.

Paul: I'd like to discuss magnetism. I searched the scientific literature but was unable to find a reasonable explanation for the forces between magnets. To me, such forces require a medium that is affected by the magnets. I have proposed that the medium, aether, for light transfer is a concentrated matrix of protons and unpaired electrons.

Magnetic properties of materials are believed to be associated with alignment of some of their unpaired electrons. A magnet would be expected to cause some of the aether electrons in its vicinity to align. A second magnet would, of course, affect the aether in its vicinity similarly. The magnets would be expected to attract or repel each other through the aether depending on their relative orientations. This concept also leads to very simple explanations for electric motors, dynamos, etc.

Maxwell: I used to wonder how a rotor, which is separated from the rest of the motor by air, could turn with such power that I could not stop it, with my hands. At that time, I considered that as further indication of an invisible material between the rotor and the rest of the motor.

Paul: I won't embarrass you by asking what you believe now. I am convinced that many experimenters have obtained hydrogen in and from vacuum. If there is a medium for light, it must extend as far as a telescope can see. I suspect that knowable space is a continuous matrix of protons and unpaired electrons. Skinner's preparation of hydrogen by electrolysis of vacuum tends to confirm this.

Maxwell: The weights of protons and electrons are known and we have a fair idea of their diameters. Your proposed medium would be extremely dense.

Paul: That's correct, even though today's scientists are searching for something they refer to as dark matter to account for the stability of galaxies, they are not ready to accept such a medium. They calculate that such dark matter must comprise at least 90 % the mass of the universe. My proposed medium would constitute over 99 % of the mass of the universe.

Maxwell: Doesn't that bother you?

Paul: It did but I got used to it. Human beings have always grossly underestimated the size and weight of the earth and the universe. It may be because the larger the universe: the smaller we feel. On what other basis does the layman decide on the size or mass of the universe?

Maxwell: That's an interesting point of view. Let me change the subject. How do you explain the permittivity or, in another words, dielectric constant, of vacuum?

Paul: My answer requires me to address the audience. (To the audience.) Don't worry. This is relatively simple, if you accept my aether. First, let's get an idea of relative size in the particle world. Atoms are extremely small even when compared with the smallest object that can be seen with an optical microscope. I searched the literature to get an idea of the relative sizes of atoms, nuclei of atoms, protons and electrons and developed a new standard of length. It is the classical diameter of the electron. I named it the audience or a. If I can get science to accept these ideas, you will famous. I am going to shrink us all down to a size where we can observe these particles. Don't be upset. Think what you are going to save on food. You may be surprised at what I am going to tell you. It certainly surprised me and I'm supposed to know this stuff.

At your new stature an electron appears to have a diameter of a dime. That is 1 audience. (a.)

A proton is similar in size. The nucleus of an average atom has a diameter of about 3.3 dimes, or 3.3 audiences. Here's the surprise. An average atom has a diameter of 33,000 dimes or 33,000 audiences. That means that the diameter of an electron is to that of a dime as the diameter of an atom is to the length of 6 football fields. There is plenty of room between the atomic nuclei of solids for great multitudes of protons and electrons. In other words, materials, which are combinations of atoms, are open sieves to the aether, I have proposed. If you could, somehow, remove the aether from an evacuated bell jar it would quickly refill with aether from outside the bell jar. It would be similar to pumping water from a cylindrical screen when the screen is immersed in a lake. This agrees with

Huygens' conclusion, earlier in this dream, that the medium for light flows through mercury and/or glass.

Maxwell: That's very interesting. What has it to do with dielectric constant.

Paul: I'm trying to develop a background to make the concept easier.

Maxwell: Sorry. Go on.

Paul: Flashlight batteries contain reactive chemicals. In reacting they cause electrons to be removed from one of the electrodes, the anode, and amassed on the other electrode, the cathode. A typical cell produces a voltage difference between the anode and cathode of 1.5 volts. At this point the reaction stops. If a copper wire connects the anode and cathode electrons will flow through the wire from the cathode to the anode until the chemical reaction is completed and the battery is "dead".

Now, let us attach a wire from one electrode of a fresh battery to a metal plate and a second wire from the other electrode to a parallel metal plate. Electrons will build up on the plate attached to the negative electrode and be removed from the other plate, until the voltage across the plates is 1.5 volts. The parallel plates are a form of capacitor. A capacitor stores electricity. If we fill the space between the plates with a material, we will find that the capacitor will store more electricity, at the same voltage.

The relative dielectric constant of a material is defined as the ratio of the amount of electricity (or number of electrons) stored by a capacitor when filled with the material to the amount stored when it is filled with vacuum. A capacitor is believed to work by the following mechanism:

The extra electrons on the negative plate of the capacitor repel negative charges in the material, while the net positive charges in the other plate of the capacitor attract negative charges in the material. This causes an electrical distortion in the material that tends to reduce the voltage across the capacitor and permit more electrons to flow from the negative electrode of the battery to the capacitor and from the positive plate of the capacitor to the positive electrode of the battery. For this reason, a capacitor filled with a material with a relative dielectric constant of two will store twice as much electricity as a capacitor filled with vacuum, which, by definition, has a relative dielectric constant of one. That vacuum is capable of storing electricity suggests, to me, that vacuum contains positive and negative charges. This is expected if the space between the plates of the capacitor is filled with a matrix of protons and electrons.

Maxwell: Why shouldn't void be able to store electricity?

Paul: Because void contains no charged particles. Wouldn't you expect air, which contains particles made up of protons and electrons, to have an infinitely greater dielectric constant than void? The dielectric constants of vacuum and air are practically the same and much greater than would be expected of air.

I have been talking about relative dielectric constant because it is a simpler concept than dielectric constant itself. Your equations require that vacuum have definite values of dielectric constant and permeability. How can void have such properties?

Maxwell: I must plead the Fifth Amendment. (A gentleman enters at the back of the stage. He stacks some of the cards to form the base of a pyramid.)

Paul: That is what I expected. Let's discuss the effects of alternating current on relative dielectric constant. With normal line current, where the direction of the current reverses 60 times per second, the dielectric constant of most materials is the same as with direct current, where the direction of the current does not change. This is true even at frequencies as high as a billion cycles per second. As the frequency is increased further, many of the effects that contribute to dielectric constant (for example the movement of atoms or ions) are too slow to respond, due to inertia. This causes the dielectric constants of materials to decrease dramatically. At the much higher light frequencies, only electrons have low enough inertia to respond. I consider the observation that the dielectric constant of vacuum does not change with frequency as further evidence of the presence of electrons in vacuum.

Maxwell: Much of your latest statement includes information that was not available to me. I don't think that I should comment. (The gentleman at the back walks up to Maxwell.)

Gentleman: Hello James. I've been sent to replace you for a while. How goes it?

Maxwell: Nice to see you Neils. Meet Paul. He has very interesting ideas. I'm afraid that he thinks more like me than like you. Be ready for an interesting discussion. (To Paul). Give him hell, Paul. It has been fun talking with someone with 100 year old opinions.

Paul: (As Maxwell leaves.) If you had lived to be 200 years old, you could be a great help to me. Thanks for listening.

Gentleman: Hello, Paul. It sounds as if I am going to have to watch out for you.

Paul: And visa versa. Welcome to my dream. (To audience) I'd like to introduce Professor Neils Bohr, from Denmark. Professor Bohr is a Nobel Lauriat. He worked under Sir J.J. Thomson and, then, Lord Rutherford, in England before returning to Denmark and is best noted for his structure of atoms. His structure of the hydrogen atom explained the spectral lines of the hydrogen atom. He assumed the electron rotated around the central proton in orbits and proposed that only certain orbits were available. As an electron transferred from a higher energy orbit to a lower energy orbit, a fixed amount of energy was given off in the form of electromagnetic radiation of the appropriate frequency. Each frequency in the hydrogen spectra could be calculated from such a transfer. This was the first important contribution to the field of quantum mechanics.

Bohr: That is a good summary of some of my work. By the way, I hope you don't mind. While you were talking with Professor Maxwell, I stacked some of the cards back there.

Paul: That's ok. That's why they are there. I am highly impressed with your work and have no quarrel with your brilliant mathematics. It seems to me, however, that physicists of your era pointed out that the atom of your model would not be stable. The electron rotating around the proton was expected to lose energy and eventually fall into and combine with the proton at the center.

Bohr: Yes. That kept coming up. It was a good argument that I could not refute. However, as my mathematics kept explaining more and more actual observations, the argument practically disappeared. After all, it is most important that science progresses.

Paul: I have no argument with your mathematics, but correct mathematics may be subject to more than one interpretation. Just because one can't think of another interpretation, does not mean that the one you can think of is correct. In the middle ages, scientists agreed that the earth was the center of their universe. They had proved it mathematically. When Copernicus showed that the movement of planets could be explained much more simply with a sun centered system, his ideas were rejected. Scientists were able to complicate their mathematics and thus retain their preconceived notions. This did not speed the advancement of science.

Bohr. Do you have an alternate suggestion?

Paul: Yes, but I'm not sure that it is better than yours.

Bohr. I'd like to hear it.

Paul: Here goes. When you were stacking cards at the back of the stage, you may have heard some of my conversation with Professor Maxwell.

Bohr. I couldn't help but eavesdrop.

Paul: Your one time boss, Lord Rutherford, was the first to demonstrate that the nucleus of an atom is extremely tiny compared to the atom itself. He found that speeding alpha particles (positively charged helium nuclei) easily passed through solids. Only a very tiny percentage of the helium nuclei were scattered back by the positive nuclei that made up the solids. This indicated that the space between nuclei of solids is amazingly larger than had previously been expected. As I pointed out to Professor Maxwell, the spaces are enormous compared to the size of protons and electrons and may be filled with a matrix of protons and electrons. I have been trying to convince the audience that such a matrix is the medium that carries light. Transparent materials like glass transmit light and, therefore must contain the medium that carries light. Even solids that we consider to be opaque to light contain the medium because, if they are thin enough, they transmit light.

Picture a material as being made up of widely separated nuclei surrounded by the proposed matrix or, better yet, look at the screen at the back of the stage. Consider each nucleus as a spherical positive plate of a capacitor. The aether in the vicinity of a nucleus would be distorted in such a way that aether electrons are nearer the nucleus and aether protons farther from the nucleus. The distortion would be less and less as the distance from the nucleus increased. The overall effect would be a decrease in repulsion between adjacent nuclei with separation. At some relatively large separation, the repulsive force may be balanced by an attractive force, for example, gravity.

Bohr. That explains the stability of the atom because there is no orbital motion involved. It doesn't explain the separate spectral lines of hydrogen.

Paul: You're right, but let me dig myself into a deeper hole. The electromagnetic spectrum includes radio, radar, infra red, visible and ultraviolet light, x-rays and gamma rays. The difference is simply one of frequency. It is believed that the effect travels as photons. But what is a photon?

Bohr: You really don't expect me to tell you.

Paul: I understand that would be a no no. There is considerable evidence that the higher the frequency of a photon, the more energy it carries and the smaller it is. As an example, x-ray photons of high enough frequency pass through crystals at some orientations and are reflected at other crystal orientations. They must be smaller than the distance between atoms. This effect is used to determine crystal structure. It is similar to a blind man throwing balls in the Parthenon, in an effort to find aisles through the columns. According to Huygens, light is passed from an activated aether particle to another aether particle. Since electrons are much smaller than x-ray photons, a photon must consist of a series of activated ether particles. Higher frequency photons are made up of fewer and more active electrons. If this is the case, the aether can transmit only frequencies (or energies) whose photons are made up of integral numbers of electrons. There are no fractional electrons. That means that there are energies that cannot be transferred through the aether. The higher the frequency, the greater the difference between possible energies. This may be the basis of quantum mechanics.

Bohr: I fear that you are confusing the audience. Can you cut this short?

Paul: I'll try. Consider the hydrogen atom. Only certain energies can be transferred to the atom through the aether. Any energy absorbed by the atom will change the arrangement of affected electrons around the nucleus and result in a more energetic atom. Schrodinger's wave equations are believed to describe the orbits of the various electrons

112

surrounding a nucleus. Instead, they may describe a contour map of the effect of the nucleus on the aether in the vicinity of the nucleus.

Bohr: Did you expect the audience to understand that?

Paul: No. I'm not sure I follow it myself, but I like it better than your explanation.

Bohr: Earlier, you suggested that the force holding atoms together might be gravity. Can you elaborate on that?

Paul: Sure, but I'm treading on very thin ice. When I discussed the size of electrons and protons, I purposely compared them to dimes rather than grapes. I picture electrons and protons of the aether as spinning disks much like coins spinning on a flat surface. Nuclei of atoms are made up of protons and neutrons.

I find the descriptions of neutrons in the scientific literature extremely confusing. On one page, I read that the neutron has a half - life of about ten minutes. It decomposes into a proton, an electron and, supposedly, an antineutrino. Its mass is slightly greater than that of a hydrogen atom, which suggests that it is, likely, similar in size to a hydrogen atom. A few pages further on, I read that a neutron is absorbed in the nucleus of an atom in about one thousandth of a second. This suggests that the neutron easily fits inside the nucleus. I suggest that the first described neutron is a combination of a proton and electron where both the proton and electron are spinning like coins on a table. The second described neutron may be a much lower energy particle where the proton and electron have lost their spins and come together like two dimes to form a much smaller combination, which fits in the nucleus. If this is the case, all atoms may be made up of protons and electrons and the aether contains everything required to form all existing materials. This leads to a simple explanation for the formation of hydrogen and, therefore, stars in space.

Bohr: Have you convinced any living scientist that this could be the case?

Paul: No.

Bohr: Do you expect to convince anybody?

Paul: No.

Bohr: Then, why do you persist in this?

Paul: I like puzzles and this is the most interesting puzzle, I have found.

Bohr: That is the response of a true scientist. By the way, you haven't explained gravity yet. (Albert Einstein enters at the back of the stage and stacks cards above the base that Bohr had assembled.)

Paul: I've been trying to build the proper background. I'm happy with the idea that magnetic forces are transferred through the electrons of the aether. I am not as happy with the idea that gravity is transferred through the protons of the ether. However, if one accepts that the neutron is a combination of a proton and an electron, the weight of any material is extremely close to the weight of the protons it contains. If one assumes that there is an attractive force between protons that remains after electrical forces are cancelled by electrons, one can come very close to explaining gravity.

Bohr: You are not happy with that description of gravity?

Paul: Not very. It needs more work.

Bohr: I see that you have a very distinguished visitor. If anyone can pick apart your suggestions, it is Professor Einstein. (To Einstein.) Hello Albert. I assume you are here to take your turn. (To Paul.) Goodbye Paul. I enjoyed your monologue. If you eventually make it to heaven, it will be my turn to do the talking. (As Bohr leaves, he pauses to admire the enlarged house of cards.)

Paul: I'm sorry. Sometimes I get carried away. I look forward to your monologue, but I'm in no real hurry. (To Einstein) It is a great honor to meet you Professor Einstein. I won't introduce you because your fame has preceded you.

Einstein: It is a pleasure to meet you Paul. You needn't be so formal. You may call me Albert. In fact, you can call me anything but wrong.

Paul: Well, Albert. I hope you don't mind if I call you a genius. I have nothing but respect for your mathematics, but I would like to discuss some of your interpretations. After all, many of them are close to 100 years old. I have the advantage of the results of many more years of research by brilliant scientists.

Einstein: To prove that I have an open mind on these matters, I will quote from a letter I wrote to a fellow professor in March of 1949:

"You can imagine that I look back on my life's work with calm satisfaction. But from nearby it looks quite different. There is not a single concept of which I am convinced that it will stand firm, and feel uncertain whether I am in general on the right track."

The rest of this play was presented earlier in this book under, "My Conversations with Einstein".

AN UNEXPECTED SOURCE OF
CLEAN ENERGY? PART 1

This is a copy of one of my papers published in Infinite Energy Issue 67, 2006.

It also can be found along with many others on Sepp Hasslberger's blog by googling: sepp paul rowe

Magnetism

How can intelligent people play with two bar magnets for five minutes and continue to believe there is nothing, but air between the magnets? Based on theories they do not understand, they are convinced that space is void and magnetism, light and gravity arc, somehow, transferred through the void.

Can void in the vicinity of a magnet have different properties than void far from the magnet? Can the properties of void near a magnet change when a second magnet approaches? If nothing, but air, separates the rotor from the rest of a motor, how can the motor develop enough power to destroy your hand if you try to stop the rotation?

In developing wave equations, both Huygens and Maxwell assumed space was filled with touching material particles. Since their equations correctly predict important properties of light, their concepts were accepted as fact, until early in the twentieth century.

Huygens[1] referred to an experiment, in which Torrecelli (A contemporary of Galileo) filled a U-tube with mercury to a sealed end and evacuated the tube through the open end. Light passed through the space that developed at the sealed end. Huygens concluded that the medium for light transfer was present in the vacuum and that the medium easily passes through the glass and/or the mercury. He proposed that the medium was made up of extremely fine, touching material particles, which transferred light by a mechanism similar to that by which sound travels through the air. He suggested that the energy is transferred much like the transfer of energy from sphere to sphere in a series of suspended metal spheres. All the energy on one sphere is transferred to an adjacent sphere. The velocity of energy transfer depends on the physical properties of the spheres.

The following is a translation from Huygens:

"And it must be known that, although the particles of the ether are not thus in straight lines, as in our row of spheres, but confusedly, so that one of them touches several others. This does not hinder them from transmitting their movement and spreading it always forward."[2]

He assumed that each activated ether particle is the start of a new wave and, on this basis, developed equations that predict observed diffractions. For many years, scientists considered this strong evidence for a material ether.

The following quotes are from Maxwell[3]

"In several parts of this treatise an attempt has been made to explain electromagnetic phenomena by means of mechanical action transmitted from one body to another by means of a medium occupying the space between them."

"According to the theory of undulation, there is a material medium which fills the space between the two bodies and it is by the action of contiguous parts of this medium that the energy is passed on from one portion to the next til it reaches the illuminated body."

1   Huygens, C. 1988. Great Books of the Western World, Vol. 34, Encyclopedia Britannica, Inc., 13th. Printing. pp. 556-560

2   Ibid. p. 560

3   Maxwell, J.C., 1891, A Treatise on Electricity and Magnetism, Vol. 2 unabridged, 3rd ed ., republished 1954, Dover Publications, New York, pp. 431-443

"Let us next determine the conditions of propagation of an electromagnetic disturbance through a uniform medium, which we shall suppose to be at rest, that is to have no motion except that which may be involved in electromagnetic disturbances. Let c be the specific conductivity of the medium, k its specific capacity and u its magnetic permeability."

Both Huygens and Maxwell based their equations on the presence of touching material particles in vacuum. When I studied physics at MIT (many years ago), the professor told the class that there is no ether, but magnetism and light are much easier to understand if one, temporarily assumes an ether. In my opinion, he was half right. I found that light and magnetism were much easier to understand, when based on a material ether.

The following four quotes are from Albert Einstein:

"In Maxwell's theory there are no material actors."[4]

"The Ether does not exist."[5]

'The electromagnetic fields are not bound down to any bearer, but are independent realities which are not reducible to anything else."[6]

"You seem to think that I look back on my life's work with serene satisfaction. Viewed more closely, however, things are not so bright. There is not an idea of which I can be certain. I am not even certain that I am on the right road."[7]

Many respected experimenters have reported the surprising appearance of hydrogen in their experiments. The following quote is from Sir J.J. Thomson:

"I would like to direct attention to the analogy between the effort just described and an everyday experience with discharge tubes. I mean the difficulty of getting the tubes free from hydrogen when the

4   Einstein, A. and Infeld L., 1938. The Evolution of Physics, republished 1951, Simon and Schuster, New York, p. 152

5   Zukav, G. 1979, The Dancers of the Wu Li Masters, William Morrow and Company, New York, p. 150. Original reference was: Einstein, A. 1920. Aether und Relativatstheorie, 1920, trans. W. Perret and G. B. Jeffery, Sidelights on Relativity, London, Methuen, 1922 (reprinted in Physical Thought from the Presocratics to the Quantum Physicist by Shmuel Sambusky, New York, Pica Press, 1975, p. 497).

6   Ibid.

7   Clark, R.W. 1971, Einstein: The Life and Times, World Publishing Co., p. 613.

test is made by a sensitive method like that of positive rays. Though you may heat the glass tube to the melting point, may dry the gases by liquid air or cooled charcoal and free the gases you let into the tube as carefully as you will from hydrogen, you will get hydrogen lines by the positive ray method, even when the bulb has been running several hours a day for nearly a year."[8]

The following is the introduction to an extremely interesting paper by Clarence A. Skinner of the University of Nebraska:

"While making an experimental study of the cathode fall of various metals in helium it was observed that no matter how carefully the gas was purified the hydrogen radiation, tested spectroscopically, persistently appeared in the cathode glow. Simultaneous with this appearance there was also a continuous increase in the gas pressure with time of discharge.

This change in gas pressure was remarkable because of its being much greater than that which had been observed under the same conditions with nitrogen, oxygen or hydrogen.

"Now the variation in the cathode fall with current density, was found to be so like that obtained earlier with hydrogen that it appeared necessary to maintain the helium free of the latter in order to make sure that the hydrogen present was not the factor causing this similarity in the results. Futile endeavors to attain this condition led to the present investigation, which locates the source of the hydrogen in the cathode, shows that the quantity of hydrogen evolved by a fresh cathode obeys Faraday's law of electrolysis, and that a fresh anode absorbs hydrogen according to the same law."[9]

Skinner employed various metals as cathode and found that most tarnished during discharge in helium. Skinner obtained thousands of times more hydrogen from a silver cathode than it could have originally contained:

"Altogether about two cubic centimeters of gas have been given off by this silver disk which is 15 mm. in diameter and about 1 mm. thick. It shows no sign of having its supply of hydrogen reduced in the least."[10]

---

8 Thomson, J.J., 1914, Nature, **90,** pp. 645-647.

9 Skinner, C.A. 1905, Phys. Rev. **, 21**, pp.1-15.

10 Ibid., p. 6.

Since the gases tested by Thomson were produced in a discharge tube, hydrogen gas may have been produced similarly, as long as he continued the discharge. I have produced surprising quantities of hydrogen gas by combusting mixtures of aluminum powder and cupric oxide in a fairly good vacuum. Mixtures containing excess aluminum produced the most hydrogen.[11]

The Ether

I am convinced that there is something in vacuum that can be converted into hydrogen and suspect that vacuum contains a concentrated matrix of protons and unpaired electrons. Such a matrix conforms to the assumptions of Huygens and Maxwell. Magnetism is generally attributed to the presence of oriented unpaired electrons. Unpaired electrons in the proposed matrix in the vicinity of a permanent magnet would tend to orient and this orientation would be considered a magnetic field. This concept permits a simple explanation for the forces between separated permanent magnets and a reasonable mechanism for the transfer of light.[12]

Helium below two degrees Absolute is a super fluid. It has zero viscosity and, once in motion, continues to flow through tightly packed granules indefinitely. Many scientists believe that it is Bose-Einstein Condensed.[13] Recently, scientists have produced Bose-Einstein condensed rubidium, potassium and sodium.[14] These condensates transfer light, but at a much lower velocity than vacuum. Could the matrix that permeates knowable space be Bose-Einstein condensed hydrogen? A matrix of touching protons and electrons having particles

11 Rowe, P.E., 1996, Hydrogen Gas from Vacuum, Part I, J. of New Energy, Vol. 1, No. 2. Pp.108-111.
12 Rowe, P.E., 2002. Light, Gravity and Einstein's Twin Paradox: An Argument for Classical Physics. Infinite Energy 7, 42, pp. 65-68. Rowe, P.E., 1998, Hydrogen from Vacuum. Infinite Energy 3, 17, pp. 80-82. Rowe, P.E., 1998., Time, Mass and Velocity. Infinite Energy, 3, 17, pp. 84-85
13 Silvera, I.F. and Walraven, J., 1982, The Stabilization of Atomic Hydrogen , Scientific American, January , pp. 66-74
14 McCook, A, 2000, The Nobel Prizes for 2001., Scientific American, December , p.29. Hau, L..V., 2001, Frozen Light, Scientific American, July, pp. 66-73. G.C.P., 2000, Hydrogen Man, the Godfather of BEC. Scientific American, December, p. 98.

of the classical diameter of the electron and masses of the proton and electron would be very massive, indeed. Substituting the calculated density into the equation developed by Bose and Einstein indicates that such a matrix would be stable at extremely high temperatures. Due to its great mass, the matrix near the earth would be attracted to the earth and tend to move with the surface of the earth. Since their equipment moved with the earth, the Michelson and Morley interferometer observations are as expected. Some of my papers[12] attempt to explain other physical phenomena based on the presence of such a matrix. If such a matrix exists, why isn't it obvious to us? How can your hand move through it with no effort? Your hand moves through it effortlessly, just as a net would move effortlessly through a zero viscosity liquid. In other words, a body in motion, through the ether, will remain in motion until a force is applied. Why aren't we crushed by the weight of the ether above us? If I am correct, the great majority of your mass is the ether within you. When you weigh yourself, you get your total weight less the weight of the ether within you. A fish weighs much less in water than in air. You don't notice the weight of the air pressing on you from all directions and, similarly you don't notice the weight of the ether. Could the dark matter for which scientist are searching be a matrix of protons and electrons that fills the knowable universe and is the transporter of light and gravity? I suspect that it is.

Possible Source of Clean Energy?

All the techniques that appear to produce hydrogen from vacuum (ether?) require the input of considerable energy. If one converted hydrogen into such an ether, one would expect to produce considerable energy. If more energy than is required to convert water into hydrogen and oxygen is produced, one could obtain clean energy from water. Using a Tesla coil that delivers a ten inch spark, in air, I have passed discharges (separately) through similar glass tubes containing hydrogen, helium and argon (pressure about one quarter of an atmosphere) for five minutes. The tube containing hydrogen became much hotter than the other tubes. This doesn't prove anything, but I find it interesting. I have neither the equipment nor the talent to carry on with this experiment. Incorporation of proper catalysts may enhance this effect.

Could some of the energy produced by lightning be produced by conversion of the hydrogen in moist air into the ether, under high voltage discharge? According to the Encyclopaedia Britannica:

"In the average thunderstorm, the energy released amounts to about 107 kilowatt hours, which is equivalent to a 20 kiloton nuclear warhead. A large, severe thunderstorm might be 10 to 100 times more energetic."[15]

Recently experimenters[16] have found that X rays and gamma rays are produced in certain portions of the lightning cycle. What is the source of this energy? Could it be from conversion of hydrogen in moist air into Bose-Einstein condensed hydrogen to produce energy and oxygen?

15   1989 Encyclopaedia Britannica, Vol. 16, pp. 472.2b.

16   Dwyer, J.R. 2005, A bolt Out of the Blue, Scientific American, May, 2005, pp. 64-71

AN UNEXPECTED SOURCE OF CLEAN ENERGY? PART 2

Introduction

Part I of this paper[1] suggested that the knowable universe is filled with a concentrated matrix of protons and electrons, possibly Bose-Einstein condensed hydrogen. Such a matrix is consistent with the medium assumed by both Huygens and Maxwell in developing their wave equations. Conversion of the hydrogen atoms in water into this matrix (the aether) would be expected to produce enormous quantities of energy and oxygen gas. Such a reaction may be the source of the energy produced in lightning storms. Perhaps high voltage discharge in the proper pressure of water vapor and in the presence of the proper catalysts would produce great excesses of energy. The paper gave a simple explanation for the forces between magnets separated by vacuum.

The present paper includes evidence that the proposed matrix would be paramagnetic and, thus, would be affected by neighboring permanent magnets. The paper also includes a brief history of Bose-Einstein condensation.

---

1   Rowe, P.E., "An Unexpected Source of Clean Energy"? Infinite Energy, issue 67, 2006, pp. 33-35

Magnetism

The following quote is from Pauling's, "Nature of the Chemical Bond"[2], for which he received his first Nobel Prize:

"The Pauli exclusion principle requires that no more than two electrons occupy a single orbit, and that the two electrons in the same orbit have opposed spins, and thus mutually neutralize their magnetic moments. The most stable orbit in every atom is the 1s orbit of the K shell. In the normal hydrogen atom this is occupied by one electron, the spin magnetic moment of which makes monatomic hydrogen gas paramagnetic. In the normal helium atom the 1s orbit is occupied by two electrons, which are required by the exclusion principal to have opposed spins; in sequence of this helium is diamagnetic, the spin magnetic moment of the two electrons neutralizing one another."

"It is customary to refer to electrons with opposed spins as paired, whether they occupy the same orbit in one atom or are involved in the formation of a bond."

I propose that space is filled with a matrix of protons and electrons and the structure may be similar to that of molten salt. Just as no chloride ion touches another chloride ion; no electron touches another electron and the electrons are not paired. Such a matrix would be paramagnetic and respond appropriately to an approaching magnet. The presence of such a matrix permits simple explanations for the forces between separated permanent magnets

Bose-Einstein Condensation (BEC)

The 2001 Nobel Prize for physics was awarded to Eric A. Cornell and Carl E. Wieman of the University of Colorado and independently to Wolfgang Ketterle of MIT for producing Bose-Einstein condensates (BEC). A group headed by Cornell and Wieman produced the condensate of rubidium in June of 1995. Ketterle's group produced the condensate of sodium in September of 1995. Each group cooled gases of the atoms to almost zero degrees absolute and produced "superatoms", which are combinations of many particles which behave

---

2   Pauling, L. "The Nature of the Chemical Bond". Cornell University Press, Ithica, NY, 1945 pp. 21-22.

like individual particles[3]. Since 1995, a Bose-Einstein condensate has been made from lithium but there was difficulty in condensing hydrogen. In 1998, a team led by Ketterle produced Bose-Einstein condensed hydrogen from spin polarized hydrogen atoms[4].

Perhaps experimenters had previously produced Bose-Einstein condensed hydrogen, but couldn't detect it, because it simply became part of the aether. It is difficult to detect water created in a lake? Ice is easy to detect.

Bose-Einstein condensed hydrogen is a combination of protons and electrons in their lowest possible energy state. Conversion of hydrogen molecules into a Bose-Einstein condensate would be expected to release considerable energy.

A 1982 "Scientific American" article by Isaac F. Silvera and Jook Walraven [5] includes an excellent description of Bose-Einstein Condensation. The following quotes are from that article:

"The statistical theory that describes atoms was first studied by the Indian physicist S. N. Bose and is called Bose statistics. The phenomenon predicted by Einstein is a mathematical consequence of Bose statistics, but it was so contrary to the intuition of physicists in the 1920's that it was then regarded as a mathematical oddity that would never be found in a real system. It is now thought, however, that the phenomenon is observable in the laboratory. It is called Bose-Einstein condensation."

"In a Bose-Einstein-condensed gas, however, a large fraction of the atoms would occupy the ground state at an experimentally accessible temperature, and nearly 100 percent of the atoms would become condensate atoms at a temperature above absolute zero."

"The most sought-after quantum phenomenon is a sudden condensation of a large proportion of the atoms in the gas into a state of minimum energy. The condensation is expected to take place at a low temperature that depends only on the density of the gas. For example, at a density of 10^{24} atoms per cubic centimeter

3   McCook, A. ," The Nobel Prizes for 2001", Scientific American, Dec. 2001, P. 29.

4   G. C. P.," Hydrogen Man, the Godfather of BEC", Scientific American, Dec. 2000, p. 98.

5   Silvera, I., & Walraven, J., "The Stabilization of Atomic Hydrogen", Scientific American, Jan., 1982, pp. 66 –75

the critical temperature is 0 .016 degrees K, whereas at the density of interstellar hydrogen the critical temperature is 10^{-18} degree K. The critical temperature for the condensation is proportional to the density raised to the 2/3 power."

"It is the coherent motion of the condensate atoms of a Bose-Einstein-condensed gas that is expected to give rise to extraordinary macroscopic properties at temperatures well above absolute zero."

"It is highly possible but not yet definitely established by experiment that superfluid helium 4 is Bose-Einstein condensed."

"Liquid helium 4 at or below 2.18 degrees is therefore called a superfluid. If it is set flowing in a tube closed on itself, the liquid continues to flow without friction, never coming to a stop as a normal fluid would. It flows into the smallest passages of its containing vessel and has the remarkable ability to flow through a densely packed powder as if the barrier were not present. A vessel with microscopic holes that would be impenetrable to a normal fluid can be as a leaky sieve to a superfluid. Such a vessel is said o have a superleak."

The above quotes led me to the following conclusions:

A Bose-Einstein condensate is a group of atoms in their lowest possible energy state.

To achieve this state considerable energy must be removed from room temperature atoms.

Production of a Bose-Einstein condensate from atoms would release considerable energy.

A Bose-Einstein condensate that was stable at elevated temperatures would be extremely dense and have interesting properties, including zero viscosity.

Light.

The July 2001 issue of Scientific American includes an article by Lene Vestergaard Hau titled, "Frozen Light"[6]. The article describes experiments her group performed at the Rowland Institute in Cambridge, MA. They passed laser beams into Bose-Einstein

---

6   Hau, L.V., "Frozen Light", Scientific American, July 2001, pp.66-73.

condensed sodium and found that it transferred light at a much lower speed than vacuum or any other known material. In fact, they were able to stop light transmission and then restart it at will, using appropriate laser beams. This led me to assume the following:

Bose-Einstein condensed sodium transfers light. There is a medium in vacuum that transfers light.

Many experimenters have reported the appearance of hydrogen gas in vacuum[7]. This suggests the presence of protons and electrons in vacuum and that the medium may be Bose-Einstein condensed hydrogen.

In order to be stable at elevated temperatures, the Bose- Einstein condensed hydrogen must be extremely dense.

We do not perceive the presence of such a medium since it has no viscosity.

The medium permeates all materials and, since we weigh materials by difference, we don't appreciate the density of the medium, which may be the dark matter required to explain the stability of galaxies.

In spite the successes of Huygens' and Maxwell's equations for the behavior of light, based on the presence of such a medium, theoretical physics is based on on the absence of any medium.

More on Magnetism

One of my paper[8] attempts to explain many observed phenomena by assuming vacuum contains a concentrated matrix of protons and electrons. As mentioned earlier such a combination might be stable, just as molten sodium chloride is stable. Sodium chloride is a concentrated matrix of positive sodium ions and negative chloride ions. Could Bose-Einstein condensed hydrogen be a similar matrix of protons and electrons? Production of gaseous hydrogen from vacuum requires the input of considerable energy. Production of

7   Rowe, P. E.," Hydrogen Gas from Vacuum, Parts I & 2", J. of New Energy, v. 1, no. 2, Summer 1996, pp. 108-115.

8   Rowe, P E., " Light, Gravity and Einstein's Twin Paradox. An Argument for Classical Physics", Infinite Energy, issue 42, 2002, pp. 65-68

Bose-Einstein condensed hydrogen from hydrogen gas is expected to release considerable energy.

A permanent magnet is believed to contain unpaired electrons some of which are oriented in a preferred direction. Heating the permanent magnets above their Curie temperature allows these electrons to orient randomly and removes the permanent magnetism. The former magnet is still attracted to a permanent magnet, because some of its unpaired electrons will be aligned by the oriented electrons in the permanent magnet.

A DC current in a wire orients the needle in a magnetic compass in its vicinity. This suggests that there are oriented electrons in the current carrying wire. The greater the current, the stronger the attraction. This indicates that the greater the current, the greater the concentration of oriented electrons in the wire. When the current is stopped, the wire loses its magnetic properties. These effects are noted even when the compass and the wire are separated by vacuum.

How can the oriented electrons in one material affect the orientation of electrons in a distant material? Most scientists would reply, "The oriented electrons produce a magnetic field and the field affects unpaired electrons in a distant paramagnetic object".

Perhaps, this works mathematically, but I find such a concept unreasonable. What are fields in a void? I can't accept that empty space has different properties in the vicinity of a magnet than in the absence of a magnet and that the void at one end of the magnet is different than the void at the other end.

Magnetic affects at a distance are easy to understand, if one accepts that space includes a concentration of unpaired electrons. The oriented electrons in a permanent magnet or in a conducting wire would be expected to cause unpaired electrons in the proposed matrix to orient and, in turn, orient unpaired electrons in a nearby object. Two separated permanent magnets would be expected to attract or repel each other depending on the orientations of the matrix electrons in their vicinities.

I picture the magnetic lines of force proposed by Faraday as matrix electrons aligned by oriented non-paired electrons in a magnet. The less the distance between two magnets, the greater the attraction or repulsion.

The following quote is from Francis Bitter[9], who was, at the time, Professor of Physics at MIT:

"An electric current passing through a coil of wire will produce a magnetic field at its center. The greater the current, the stronger the magnetic field. There is no saturation effect here. So far as we know, this increase of field with increasing current continues indefinitely."

Magnetic materials reach saturation in magnetic properties when all of their unpaired electrons are similarly aligned. The above quote suggests that vacuum contains an extremely high concentration of unpaired electrons.

Conclusions:

Converting hydrogen into a Bose-Einstein condensate should result in the release of considerable energy. Such a state may be the medium for light transfer accepted by the scientific community, prior to the twentieth century. The presence of such a medium permits simple explanations for otherwise, complicated phenomena.

The enormous energy produced in lightning storms may be from the conversion of hydrogen atoms in atmospheric water into Bose-Einstein condensed hydrogen under high voltage discharge.

A serious investigation of high voltage electrical discharge in moist atmospheres may lead to the practical production of enormous quantities of cheap clean energy.

Edited; September 11, 2006

The following was added on March 17, 2008.

Could the following experiments involve conversion of hydrogen gas into the aether?

Collie, J. N., F.R.S., Patterson, H. S., Masson, I., Proc. Roy. Soc., (A), **91**, pp. 30-45 (1914) and pp. 32 and 33

Collie and Patterson surrounded a discharge tube with a highly evacuated glass tube. The discharge tube contained hydrogen. The coil used would give a 12 inch spark in air.

9   Bitter, F., Magnets, the Education of a Physicist, Doubleday Anchor Books Doubleday & Company, Inc., Garden City, NY, 1959. p. 102

The Following quotes are from pages 32 and 33:

"It was designed in order that the wires connected to the electrodes in the inner tube A passed through the outer tube B so that there were no live wires in the outer vessel. Helium and Neon were found in the outer vessel. But the remarkable fact was noticed that the hydrogen (4.6 c.c.) admitted to the inner tube and sparked, at the end of the experiment had diminished to about 0.4 c.c. Moreover, after breaking up the tube and melting the electrodes and the powdered ends of the inner tube in a hard glass tube, only 0.6 c.c. of hydrogen was obtained. This apparent disappearance of hydrogen is always a noticeable fact during the discharge and, up to the present time has not been entirely explained".

"The absorption of gases in vacuum tubes has been noticed by several experimenters:

S. E. Hill, 'Phys. Soc. Lond. Proc.,' Dec.,1912, p. 35 finds that hydrogen is absorbed in electrodeless tubes. He refers to Willows,('Phil. Mag.,' April, 1901), and Campbell Swinton ('Roy. Soc. Proc.,'A, vol. 79 (1907)"

Masson performed an arc discharge experiment in an inverted silica U-tube. The ends of the tube were in mercury. The voltage was 110 and the arc was started by raising the mercury level in the tube.

The following quote is from page 38, of the same article:
"From the first, however, the arc was protected from the air by being water jacketed, and on continued discharge no helium or neon was produced. An experiment was made in which hydrogen was admitted without affecting this result; it was noticed, however, that after about 10 minutes of passing the arc discharge the whole of the hydrogen (about 1 c.c.) had completely vanished, and the mercury could be driven to the top of the capillary."

The hydrogen may have reacted with the mercury to form a hydride. In my experiments, the hydride forms an easily observed skim on the mercury surface. Perhaps this experiment should be repeated using a greater quantity of hydrogen and a longer discharge time.

Paul E. Rowe

The above is a re-edited version of a paper published in the May-June, 1996, #8 issue of Infinite Energy.

While searching the scientific literature, I came across many interesting articles. They led me to write the following paper:

CONTROLLED TRANSMUTATION OF ELEMENTS UNDER SURPRISINGLY MILD CONDITIONS?

Introduction

Bockris and Minevski of Texas A&M University recently reported experimental results that strongly suggest that palladium atoms near the surface of hydrogen atom-saturated catalysts are transformed into atoms of other elements under certain mild electrical conditions.

Minzuno, Ohmori and Enyo reported similar results with palladium catalysts saturated with deuterium atoms. The isotope make up of the elements produced were quite different than that which occurs in nature.

It is extremely difficult (perhaps impossible) to explain these experimental results, unless one accepts that transmutation has occurred.

This article will attempt to demonstrate that experimental results reported prior to 1930 lead to a similar conclusion.

Discussion

Between 1910 and 1930, many experimenters (some extremely well respected) reported the mysterious appearance of hydrogen, helium and neon in electrical discharge tubes. E.C.C. Baly, a Fellow of the Royal Society, summarized pertinent results in the Annual Reports of the Chemical Society for 1924 (pages 41 to 47) and 1920 (pages 27 to 35). He published results of his own experiments with R.W. Riding in 1925 and 1926. They concluded that nitrogen atoms had been converted into helium and neon during their high voltage electrical discharge experiments.

On February 13, 1914, Professor J. Norman Collie, Fellow of the Royal Society, presented a speech[1] to the society. He described several experiments he performed and those reported by others in which hydrogen, helium and neon gases mysteriously appeared in electrical discharge tubes. The last paragraph of his speech follows:

"This fact cannot be explained by air leakage, for air contains four or five times as much neon as helium. At present the investigation was only begun; many more experiments would have to be made for the source of the helium and neon was still obscure; but if it could be proved that these gases were produced from many metals and other substances under the influence of cathode discharge, it is obvious that it would be a discovery of the most far-reaching importance."[2]

These experiments were later described in greater detail.

In 1923, Baly performed many variations of Collie's experiments with R.A. Bailey[3]. They found neither helium nor neon.

In a short paper, Robert Goddard, the pioneer rocket scientist, discussed the apparent production of rare gases in discharge experiments. The final paragraph of the paper follows:

"It is by no means certain, however, that the action in question consists simply in the liberation of absorbed gases, for Sir J.J. Thomson has discovered evidence of genuine production of helium and X_3 from elements (lead) and chemical compounds (salts of sodium and potassium) which suggests an actual atomic change, if not a genuine disintegration. The whole problem is very complicated, and it is the writers purpose merely to call attention to the importance of surface conditions in the production of rare gases."[4]

---

1 The Royal Institution Library of Sciences, Vol. 7, ed. Sir Wm. Bragg and Prof. Geo. Porter, Elsevier Publish Co. Ltd (1970) pp. 402-4
2 ibid, p 404
3 Annual Reports of the Chemical Soc., 1914, pp. 41-7; 1920, pp. 27-35
4 Goddard, R.H., Science, 14, pp. 682-4, 1915

X_3 was later found to be H_3 (triatomic hydrogen), which has a half-life of about one minute.[5]

Sir J.J. Thomson was awarded the Nobel Prize in physics and is known as the discoverer of the electron. In a book originally published in 1913, he described the production of helium and neon during the bombardment of various chemicals with cathode rays. The following quotes are from this book:[6]

"To test whether this was the source of the helium I bombarded soluble salts such as LiCl, NaCl, KCl, KI, RbCl, AgNO3 which were dissolved in water and also in some cases in alcohol and then evaporated to dryness, the process being in some cases repeated several times. Salts which had been treated in this way yielded helium and in some cases neon; the yield of helium from these salts of the alkali metals and in particular from potassium was exceptionally large, KI giving a larger supply than any other of the substances I examined, with the exception of those like monazite sand which are known to contain large supplies of helium. Some of the salts have yielded apparently undiminished supplies of helium, after being dissolved and evaporated ten or twelve times."[7]

"The aluminum cathode in the tube used to bombard the substances with cathode rays might be suspected as a source of helium. If this were the case, however, the rate of production of helium would not depend upon the nature of the salt bombarded, nor would it make any difference as to whether the cathode rays hit the salt or not. As both these conditions have a great influence on the rate of production of helium we may regard this source as eliminated. In addition to the proceeding considerations some of the cathodes have been in almost continuous use for months without any perceptible diminution in the rate of supply of helium."[8]

If the helium and neon had diffused through the glass walls of the apparatus, the nature of the salt bombarded would have had no effect on the rare gas concentration.

---

5   Wendt, G.L. and Landauer, R.S. JACS., 42, pp. 930-46, 1920

6   Rays of Positive Electricity and Their Application to Chemical Analysis, Thomson, J.J. 1913, (Reprinted, 1988, by Borderland Sciences, P.O. Box 429, Garberville, CA. 95440

7   ibid. pp. 124-5

8   ibid. pp. 125-6

"The view that helium can be got out of other chemical elements raises questions of such a fundamental character that few will be prepared to accept it until every other explanation has been shown untenable. It would greatly strengthen the proof if we could detect the parts of the atom which remain when the helium is split off."[9]

Perhaps, some of the atoms recently detected[10,11] are parts of atoms that remain when helium is split off.

In 1925, Riding and Baly[12] reported results of experiments performed in a discharge tube having a concave cathode, which focused onto a hollow aluminum anticathode coated with a thin nitride film. A discharge in low-pressure oxygen (below 2 torr.) produced measurable helium after 42 hours. The following quotes are from their article:

"At that time the suggestion was made to us by Prof. Masson that the results obtained by him in conjunction with Collie and Patterson might be due to the disintegration of nitrogen, this element being present in the form of nitride as a surface film on the electrodes." [13]

"The first experiments were carried out with hydrogen, but negative results were obtained. On replacing the hydrogen with oxygen a positive result was at once obtained, since after 42 hours discharge a small quantity of helium was formed. Emphasis must be laid on the fact that many previous experiments with the same induction coil and the same design of discharge tube, without a nitride layer, no trace of rare gas was ever obtained" [14]

"On repetition of this experiment, the discharge being passed for longer periods (90 to 100 hours), it was found that hydrogen, helium and neon were formed. It is of considerable interest to note that the relative proportions of helium and neon are the same as those obtained by Collie and Patterson."[14]

"We would again emphasize the fact that no trace of either helium or neon is formed in the absence of nitride film, and this would seem to exclude the presence of an air leak, since it is not possible to believe

---

9   ibid. p. 128

10,11,12   Bockris, J.O'M. and Minevski, Z., Infinite Energy, Vol. 5&6 pp. 67-9, 1996; Mizuno, T., Ohmori, T. and Enyo, M., Infinite Energy, Vol. 7, pp. 10-13, 1996; Riding, R.W., and Baly, E.C.C., F.R.S., Proc. Roy. Soc., 104A, pp.186-93, 1925

13   ibid. p. 187

in a leak which only occurs when the anticathode is coated with nitride. In addition to this, we have many times confirmed the well-known fact that the spectrum of the residual gas left after treatment of air with charcoal cooled in liquid air shows only the spectrum of neon. These results would suggest that the rare gases found by Collie and his coworkers were due to the atomic disintegration of nitrogen and not to synthesis from hydrogen. This conclusion has been confirmed by a number of subsequent experiments which may be briefly described."[14]

Other experimenters have shown that helium and neon from the air slowly diffuse through very hot glass walls into vacuum. The reasons for this view were summarized by Robert W. Lawson of the University of Sheffield. However, this does not explain the appearance of these gases only when oxygen is present and the anticathode has a nitride coat.

The following Quotes are from Baly and Riding's 1926 paper [15]:

"We have now carried out some further experiments which would seem to confirm our original results. In the first place, both helium and neon have been obtained by passing the discharge between a concave aluminum mirror as cathode and a magnesium anode through a mixture of oxygen and nitrogen under reduced pressure. No trace of the rare gases was formed if the discharge were passed in the opposite direction; and since in this series of experiments the two types of discharge were used alternatively, the same mixture of oxygen and nitrogen was used, and the apparatus was not changed in any way, this would obviate any possible criticism that the rare gases had their origin in an air leak."[16]

It also obviates any possibility that the rare gases had diffused through the walls of the apparatus.

"As regards to the origin of these gases, we believe that they arise from the disintegration of the nitrogen atom. Attention may be directed to the fact that hydrogen is always to be found in the residual gases along with the helium and neon, although the greatest possible care was taken to remove it from the electrodes before each experiment. At the same time it may be pointed out that in spite of

14   ibid. p. 18
15   Baly, E.C.C., and Riding, R.W., Nature, 118, pp. 615-6, 1926
16   ibid p. 625

all precautions it is impossible to secure the total absence of oxides of carbon."[17]

Baly and Riding's conclusion that nitrogen atoms are transformed into helium and neon atoms (and possibly other atoms) under their experimental conditions, did not conform to the, then, or present views of theoretical nuclear physics. They measured the emission spectra of gaseous products. They did not measure non-gaseous products. Bockris and Minervski's[10] results, confirmed by Mizuno, Ohmori and Enyo[11] suggest that, under somewhat different conditions, palladium atoms are transformed into platinum, silicon and zinc and possibly other atoms. They did not measure gaseous products.

Conclusions

The simplest conclusion from the above observations is that transmutation of elements has taken place under surprisingly mild conditions. Theoretical nuclear scientists find that such a conclusion defies one of their time-honored concepts. I, for one, am willing to consider that their concept is incorrect.

This is a re-edited version of a paper published in the May-June, 1996, #8 issue of Infinite Energy.

Paul E. Rowe March 4, 2007 02649rowepaul@comcast.net

---

17 ibid p. 626

Summary of the Book

The Discussion Section of this book includes conclusions, I reached, by assuming that the knowable universe and everything in it is permeated with a concentrated matrix of protons and electrons. It attempts to explain various observed phenomena, on this basis. The rest of the book tries to show that the assumptions are reasonable. The author expects that many of the conclusions will not be accepted by the scientific community. However, if the universe is permeated with a matrix of particles, as assumed by Huygens and Maxwell, conclusions based on the absence of such a matrix should be reconsidered. Could the proposed matrix be the Aether, Dark Matter and the Higgs Field?